WITHDRAWN

Biological Clocks:
Your Owner's Manual

N 0091857 1

NW 6425   Bios/95   £ 13.0⁻

Biological Clocks:
Your Owner's Manual

Sue Binkley

NEWMAN COLLEGE
BARTLEY GREEN
BIRMINGHAM, 32

CLASS   612.022
BARCODE   00918571
AUTHOR   Bin.

harwood academic publishers
Australia • Canada • China • France • Germany • India •
Japan • Luxembourg • Malaysia • The Netherlands • Russia •
Singapore • Switzerland • Thailand • United Kingdom

Copyright © 1997 OPA (Overseas Publishers Association) Amsterdam B.V.
Published under license under the Harwood Academic Publishers imprint,
part of The Gordon and Breach Publishing Group.

All rights reserved.

No part of this book may be reproduced or utilized in any form or by any
means, electronic or mechanical, including photocopying and recording, or
by any information storage or retrieval system, without permission in
writing from the publisher. Printed in India.

Amsteldijk 166
1st Floor
1079 LH Amsterdam
The Netherlands

British Library Cataloguing in Publication Data

Binkley, Sue
  Biological clocks : your owner's manual
  1. Biological rhythms
  I. Title
  612'.022

ISBN 90-5702-534-5

To

H. Randolph Tatem III, MD, my husband

Shelley Binkley, MD and *Kyra Tatem O'Brien*
my children

Connor Baker, my grandchild

and

Rich Zelter, the brave hero who fished my hypothermic
carcass out of the 44°F. Roaring Fork River

CONTENTS

PREFACE

This book, like Gaul, is divided into three parts.

I don't really expect people to read it beginning on page 1 and continuing to The End of the Whole Mess (chapter 14). Human Rhythms (chapter 13) probably has the information that will interest you most. Perhaps you only want to measure your own rhythm using section III. Freeruns (chapter 5) and Entrainment (chapter 4) contain the "guts" of how the rhythms work.

Section I explains biological rhythms as I have loved them for a quarter of a century. Egad! That's a long time...probably longer than many of you have been alive! For me, biological clocks are my old friends.

The sun and moon dance with the earth. The seasons perform a stately minuet. The night dances with the day. We creatures of the earth are exquisitely adapted to waltz with the sun. Imagine ice dancing, where you are the follower and your celestial partner leads, and, yet, if your hand is released, you can skate on your own. You may prefer to think of these celestial events as things that happen outside of you. After all, you probably sleep, as I do, inside in a cocoon lit by artificial lights with its silence broken by an alarm clock. Perhaps you, as I do, make an occasional night-time excursion to squint at the sky. Last week, from my doorstep, I saw a comet, fragile bride Hyakutake in a misty veil, dragging her train as she searched the Zodiac for her groom. I watched the earth cast a dirty shadow on the moon, a lunar eclipse that made "the man in the moon" look like a silver clown wearing a beanie.

Figures have been collected in one section for easy comparison. I suggest a quick look at them first off.

Section II contains the pedantic appendices—a glossary and an annotated bibliography. This *is* supposed to be a serious book, right?

Section III has instructions and blank charts for measuring your own rhythms. It's not essential to bother with this part, but even doing a little of it will give you new insight into yourself.

Measuring your own rhythm requires only simple tools—a clock or watch and a pencil. You can begin immediately by jotting a couple of times in the margin.

What time did you wake up this morning?

What time did you go to sleep last night?

I've been asked to explain what knowing about your biological clock can do for you. Could you, for example, use it to forecast the future, like astrology? Well, I can't promise you that you can do a darn thing with it. It's like your body temperature. You can measure your body temperature and figure out whether you have a fever or hypothermia, but otherwise it's just there, something your body does. So also your body clock is there, ticking along, and you can measure it. So, no tea leaves promised. This is not a self-help book; and it's not a textbook. There's a little information; maybe it will interest you. It's a subject that has consumed me for more than a quarter of a century.

ACKNOWLEDGMENTS

For their help and inspiration I wish to thank my editors Sally Cheney, Charlotte North and Lynn Coyle. I am indebted to John Burns, Dave Samples, Tom Lichty, Tabitha King, James Michener, Eva Shaderowsky and Richard Bachman.

I also wish to remember the plants and animals who gave the basic information on which the field of circadian rhythms is based.

I am grateful to the Bucks County Library in the Michener complex on Pine St. in Doylestown, Pennsylvania, in particular the staff of reference librarians—Inita Rusis, Sharon Warne, Jan Dickler, Barbara Brown, Amy Wardle, David Dunlap, Roberta Yakovich, Theresa Hanas, Sandra Alford, Regina Switzer, Lynn Biddle and JoAnn Kern.

In addition, I thank Naomi Ihara and the Pitkin County Library in Aspen, Colorado.

SECTION I: BIOLOGICAL RHYTHMS

Chapter #1: Fooling Around with Time

What is time? What is a rhythm? What rhythms in our environment are important to our biology? What is it all about?

Cycles

The idea of recurring cycles has been considered with respect to all sorts of phenomena—hormone secretion, cancer chemotherapy, growth of tree rings, animal populations, precipitation, crop production, commodity prices, economic recessions and booms, ice ages, tides, sunspots, eclipses, meteor showers, earthquakes, volcanic eruptions, crime, fashion, wars.

The subject of this book, however, is biological timing, the timekeeping in living organisms, in plants, and in animals including people. "Red tides," for example, recur once a year, producing a brownish red discoloration of the water when the dinoflagellates (*Gonyaulax tamarensis*, single cell marine microorganisms) which winter on ocean bottoms, rise to the surface and produce a neurotoxin that can kill fish and even the people that eat the fish. Biological cycles may be annual, tidal, lunar, menstrual, and estrous, but the main focus of this book is on daily cycles, of which the most obvious to you is your daily sleep-wake cycle.

Table 1.1 Cycle Sampler[1]

Sunspots	22 years
Declination of the moon	18.6 years
Weather fronts	7 to 8 days
Economy	4 years
Lynx and rabbit abundance	10 years
Bamboo flowering	15–150 years
Cicada emergence	17 years
Chinch bugs and tent caterpillars	9.6 years
Red tides	Once a year

The Time of Your Life

Probably the most noticeable timing that affects you is the 24 hour alternation of night and day. You have a daily cycle of sleeping and waking that is innate, it does not require light and dark to recur. It may be possible to manipulate the timing of your behavior to maximize the time available, either by altering the quality of use of the time, or by increasing the waking time.

You may think how much time you have may be limited by your lifespan. But look at this differently. Think of the time you have as the time you have awake.

If, for example, you could add 1 hour a day to your awake time, over a lifespan of 70 years, you would gain an astonishing 25,550 hours of time—for work, for leisure activities, for intellectual enrichment, or for rose smelling.

Biological Watch

You may be shocked, and certainly I was shocked when I published my first book, which was about the pineal gland, to learn that the author may not get to choose the book title. So much for the author as God. For a past book, I did manage to fight off *Feathered Clocks and Furry Calendars*. A reviewer twitted me about the title: "Using *The Clockwork Sparrow* as a title is a good example of how the author does not understand marketing. It's a good joke for insiders but has no place in marketing a textbook." Like most "reviewers," he didn't offer an alternative. Heck, I wanted to call that book, *The Biological Watch*. But my editor was right. More books were probably sold by clockwork sparrows than would have been sold by biological watches.

So, why did I want to call a book *The Biological Watch*? Living things, possess an internal chronometer. We are like other living things in that we also possess an internal clock. We carry it around with us, and therefore, to me, it seemed appropriate to call it a "biological watch." The phrase usually used is biological clock. But "clock" makes me think of something mechanical that hangs on a wall.

You might have the idea that I bounded out of my pink bassinet determined to devote my life to studying biological clocks, running around with a magnifying glass, examining ants. I admit that my interest in rhythms and light has been life long. I "spot" an early exposure to the concept of cycles, in a high school science report that I wrote on sunspots—sunspots have cycles of eleven years. In college, I was intrigued by a book about Stonehenge that said the Heelstone points to the midsummer sunrise. I did biology here and there and ended up at the University of Texas in Austin soon after a sniper had aimed his gun from the tower that was the library—proving, if you have any doubts, that libraries are dangerous.

The biology building was under the long-horned shadow of that dark tower. The ecologist already had too many graduate students, and graduate students told me that some other professors didn't want women students, so why risk being pinched by crayfish? Biological clocks, light, and the pineal gland were amusing. The professor accepted me.[2] And what else was there to do anyway?

I was not the first to be interested in rhythms. The earliest notice of the rhythm that governs man (and woman!) has been traced back more than 2500 years to Archilochus, a Greek poet. Perhaps the yin-yang nature of a thing—alternation, water and coldness versus fire and hotness and brightness—of the ancient Chinese contained the idea of cycling.[3] In 1779, D. C. Hufeland noticed the natural chronology of 24-hour time periods in our lives. J. J. Virey, a French medical student, wrote a thesis in 1814 and used the phrase "horloge vivante," which means "living clock" for daily rhythms. The first monograph about biological clocks was published by Erwin Bünning in 1958.[4] I was 12 years old, had never heard of Bünning and was probably reading Brontë's *Jane Eyre* or *Miss Pickerell Goes to Mars*.[5]

Set in Stonehenge

We have to talk a little bit about the motions of the sun and moon.

I will do it here, and for fun, include some speculations about Stonehenge.

Reckoning time has been important to people as far back along that arrow into the past as we are able to peer with our archaeological glasses. My personal favorite in the prehistory of time department is the popular Stonehenge. I have been drawn to Stonehenge the way

Neary was obsessed with the Devil's Tower in *Close Encounters of the Third Kind*. I always thought those five tones sounded like the first five notes of a song about wishing on stars. By the way, did you know that Stephen Spielberg wrote the book on which the movie is based? But, I digress.

Stonehenge, in case you were sleeping during astroarchaeology class, is a tumble of large stones on Salisbury Plain in southwestern England. I have been to visit it three times. The stones lie on a barren plain—well, now they lie in a patch of mowed grass. Viewed from any side, they appear to stand almost on the line of the horizon. We know little about them, because the architect left no written plans.

It has been great fun to speculate on the meaning of Stonehenge, a sport in which I have participated myself. We can only speculate about the solution to the mystery, but, heck, why not?

What evidence do we have? We know the arrangement of stones and holes. And there is some idea of when parts were constructed— Heel stone and Aubrey holes and Station Stones (c2800 BC), Sarsen Circle and Sarsen Horseshoe (c2100 BC) and after that, the Y and Z holes were dug.[6]

Stonehenge is arranged like a lollipop in which the candy is a series of concentric rings. The outermost ring is an outer bank. Inside that, rings in sequence are a ditch, inner bank, 56 X (Aubrey) holes, 30 Y holes, 29 Z holes, 30 Sarsen Circle of stones, and 5 three-stone trilithons that form a Horseshoe. The outer bank is nearly a hundred meters in diameter; the Sarsen Circle is about 100 feet in diameter.

The "stick" of the lollipop is called the causeway and it contains a stone, the Heelstone. The stick points to the direction of a mid-summer sunrise. Today a modern highway slices the causeway at an angle, just outside the Heelstone, and Stonehenge itself is surrounded by rope barriers and chain link fences.

Many good men have proposed that Stonehenge is an astroar-chaeological computer that marks the azimuths[7] of sun and moon positions.[8] So Stonehenge has long been visualized as a kind of "sundial."

What would have been important to people five millennia ago? Timing of events, which are important enough for us to wear watches and keep calendars, would have been important then as well. Key timed events they would surely have noticed would have been female fertility cycles, phases of the moon, and seasonal motions of the sun. Priestess scientists of Stonehenge need only to have been able to mark the days between easily observed phenomena—menses, lunar phases, solstices—to keep track of the timing of these key events.

The phallic Heelstone and the Aubrey holes were placed in the first stage of Stonehenge construction, in 2800 B.C. The 56 Aubrey holes can be used to keep track of a 28 day cycle. The number of Aubrey holes is 56, which is twice 28, which is the mode value of the length of the human female menstrual cycle. By counting both day-times and night-times, a 28-day menstrual cycle can be measured. Or, alternatively, the holes might have been used for pregnancy testing. If Boadacia began counting a hole per day beginning at the menses, then the next menses was due halfway around the circle. If, however, the counter completed a circle without another menses, then a pregnancy was likely.

The Station Stones, which may also have been placed in phase I of Stonehenge construction, could be used to predict the fertile period. Counting an Aubrey hole each day either direction from the avenue marked by the Heelstone, the Station Stones are located between Aubrey holes 9–10 and 16–18; days 9–18 would bracket the most fertile period of the menstrual cycle.

Next to the daily motion of the sun, the most obvious moving heavenly body is the moon whose phases change in a 29.5 day cycle. How can the half day be compensated? The Y and Z holes (placed in phase II of Stonehenge building, 2100 B.C.) could have been used. Count one lunar cycle with 30 Y holes, and the next lunar cycle with 29 Z holes, and continue alternating between counting Y and Z holes to predict the 29.5 day cycle.

Still, the Sarsen Circle, the most prominent part of Stonehenge, has 30 upright stones defining 30 spaces between them casting thirty bars of light and shadow. Why thirty I wondered?

The 30 stones of the Sarsen Circle and the five pi shaped Horseshoe Trilithons are the most visible features of Stonehenge and were surely meant to commemorate a grand achievement, I thought—the discovery of the circle, the wheel, or, perhaps the number of days in a year. Dividing the year into lunar cycles is partially satisfying because there are 12 lunar cycles between summer solstices. But the 12 lunar cycles only take 354 days, so counting lunar cycles to predict solstices would soon result in discrepancies. Counting 12 times 30 (the number of Sarsens), 360 days, is closer to the length of a year, but still off by a frustrating five days. However, if the priestesses[9] counted 12 times the 30 spaces between sarsens of the circle, and one times the five spaces framed by the trilithons, the year lasted 365 days!

Recently, I was pleased to read that in 2000 B.C. the Assyrians, whose empire was east of the Mediterranean and included Egypt, had a 360 day year with twelve 30-day months to which they added 15 days every third year.[10] This seemed to me the same kind of thinking I had imagined for Stonehenge.

One wonders, I did anyway, why they didn't just make smaller, pocket-sized Stonehenges to calculate menstrual cycles, estimate fertility, keep track of the cycles of the sun and moon, and point directions. Why did they have to build such a big monument when a little hand held Stonehenge would do? Who knows?[11] Monoliths last, at least until the stones are ravished to build something else. And maybe the Hengers had some good reason to walk around in it. Perhaps they measured shadow positions. Perhaps they walked around the tops of those lintels.

And it would be visible from the air. From the air, Stonehenge casts shadows much like the gnomen of a sundial. In fact, it would make a "noon mark" on the planet. What's a noon mark? Noon marks on old houses were lines drawn on floors or window sills or walls to mark where the shadow of an upright object—a flagpole, the side of the window, a doorjamb, a tripod—fell at local noon.[12]

The earth rotates every 24 hours, there are 29.5 days between full moons, and there are 365 days a year, more or less. Days, lunar cycles, and years are the numbers on the earth's clocks and calendars.

What may astonish you, is that living organisms do not need either a small-scale or full-sized Stonehenge to tell what time of year it is. They can do it by measuring photoperiod and using their internal chronometers.

Soap Box

One of the dubious privileges of authordom is that Lizzie Borden gets to grind her own particular ax. I have one edge to whett that pertains to the material herein. I have been uneasy with the rush to classify human behaviors as mental "disorders" which suddenly require expensive drugs and treatments.

This is a story without any moral. Isn't that a relief? If there is a message herein, derived from my quarter century of barrel rolling in circadian rhythms, it is that there is a wide range of biological variability, and if one happens to be a special person standing alone upon the edge of the standard curve rather than crowded with everybody else into the middle peak, that is not cause for alarm, but rather, reason to rejoice in the difference.

Chapter #2: More Seriously

Let me begin, as Herman Melville began, by dusting off the lexicons and grammars. There are words I will use. Some of these words already have a great many connotations and you and I may assign different meanings to them.

Time, for example, is not just the name of a magazine or *emit* spelled backwards. Perhaps it is worth taking some time, for the consideration of, well, *time*. Consulting a dictionary produces a, well, time consuming, consideration. In these researches, I myself learned something new, that the word *time* means "pub closing time" to some people.

Dictionaries have pages to say about the meaning of the word *time*. Time is a continuum. It is nonspatial. Events proceed in succession from the past points to the future. Time is apparently irreversible. Time is a duration. Time is a number. The number represents years, days, or minutes or a point in the continuum of time.

When I began my first class on this subject, the first question that was asked by a student was, "What is time?" Yikes! I avoided a debate, which would have lasted, well, forever, or through the available class time anyway, and answered, as I will here, "For now, time is what is shown by that clock on the wall." It was a typical analog clock, disc-shaped, with a long black minute hand, a shorter hour hand, a red second hand, the numbers one to twelve, and 60 tic marks.

But this was an evasion. There are different ways of reckoning time, and if we are to communicate, we must know which time we are talking about.

After wallowing in Webster, Hackluyt, and Richardson's dictionaries, Melville sent his Sub-Sub in search of whatever "random allusions" to whales could be found in "any book whatsoever." I tried applying Herman Melville's approach to "time" and was overwhelmed. A computer search of the Bible produced a whopping 391 verses about "time" which I will spare you. A Concordance contained a somewhat smaller list, which I will also spare you. The subject of "time" was ignored in my 1983 edition of the *Concise Columbia Encyclopedia*. Perhaps the subject of time was not "concise" enough. The index to Bartlett's quotations has a two-page exhausting list. I think I will take a cue from Scarlet O'Hara, I'll think about time tomorrow. *Webster's New Universal Unabridged Dictionary*—a dictionary from which essential, but impolite, four letter words of the vernacular have been ... abridged—had pages on the subject of the four letter word, *time*.

Most times given in this book will be Eastern Daylight Savings Time, unless otherwise noted. Years, hours, minutes, and seconds will be just what you are used to.

Poetry

Still, faced with the godzillion definitions of time, I find myself feeling that I have begged the question, that I have not described the *time* accurately. And my perception of *time* is not the only opinion. *Time* can be a slippery concept. I turn to the poets of, well, our time.

Shakespeare, of course. "Peace! count the clock!" says Brutus. "The clock hath stricken three," says Cassius. "Tis time to part," says Trebonius.[13] *Thunder and lightning!* Were there any clocks then to be stricken in Caesar's House?

Colin Fletcher wrote *The Man Who Walked Through Time*. Actually, Colin walked through the Grand Canyon contemplating geological time. He thought of time more as a pattern. All we know about time moves through this pattern.[14] That idea of *time* didn't clarify it for me. Standing on the edge of the Grand Canyon—I did not walk through it—looking at the strata of rock and the river that runs through it, one's own bit of lifetime seems microscopic.

Margaret Atwood[15] thought that time was a fluid like wind and water, not a solid, like wood. Nor did she view it, as I do, neatly subdivided into decades and centuries. She says for our purposes we have to pretend that time is cleanly divisible. Here, I will take that view, that time is something linear that we can measure and use in experiments.

Stephen King suggested that time is a melting snowman.[16] And remember Christine? Stephen King's red and white Plymouth Fury with the odometer that went backward? However intriguing going back in time may be, I can tell you I have been watching now for half a century and, except in the movies and science fiction literature, I haven't seen time go backward yet. Or back to the future. The closest one can come is to take the Concord and travel west, in which case it is possible to arrive at the destination at an earlier time than one departed, but that is only because of the arbitrary assignment of time zones. Mr. King cloned himself with pen names such as Richard Bachman (maybe to get more time?). He vanished time altogether in his story, *The Langoliers*. Several of his books bear titles that are phrases familiar to the students of biological clocks (*The Dead Zone, Night Shift*), and *The Dark Half* has the pineal gland and third eye.

Do you feel there are not enough hours in the day? The Hours were Egyptian goddesses of the underworld (Horae to the Greeks) that acted together against the enemies of the sun god, Re, who may be Osiris by night. The Hours control human destinies. They fix each person's life span. They are the order that time imposes over chaos.[17] The Horae are elsewhere said to be the seasons of cyclic death and rebirth named Dike (Justice), Eunomia (Order), and Irene (Peace).[18]

Alan Lightman wrote a delightful treatment of time in his book *Einstein's Dreams*. Each tiny chapter deals with a different view of time, time running backward, stopped, a view of the future. The chapter for 24 April 1905 compares mechanical time, where people march to the beat of their mechanical watches and clocks, and wildly magical body time in which people do things whenever they please— Alan says where the two times meet is called "desperation" and where they are permitted to express themselves separately is called "contentment."[19]

A charming fanciful treatment of time and season is the story and paintings in Kit Williams' book, *Masquerade*, the story of Jack Hare and his quest for his lost talisman, a book which has riddles and a puzzle. I won't spoil the puzzle by telling you the solution, except to say that it's so difficult that it drove me bonkers for a week before I gave up trying to figure it out, and that there is a version of the book that contains a solution.[20] I would never have figured it out.

So much for poetry and puzzles, back to biology.

Is Time the Sixth Sense?

Biologists generally recognize five senses—hearing, smell, taste, touch, and vision. Keeton explained that the phrase, the "sixth sense,"[21] has been used to denote various kinds of extrasensory perception. Perhaps humans and animals should be considered to have a "sense of time."

It might be even more appropriate to recognize multiple "senses" of time because there are so many ways in which organisms use time in their lives. In the field of biological clocks, investigators have devoted their efforts to discovering how plants and animals (including people) measure time, how they use time information, and how they organize the timing of their lifestyles and internal functions. As a result of their inquiries, we also know what happens when an organism's temporal order is disrupted, and what disrupts that temporal order.

Every day I am conscious of the rising and setting of the sun. I often notice the phase of the moon. I spent my early years in the Midwest, and had little appreciation of the tides. I was astonished when I visited Digby, Nova Scotia, which can have 40–50 foot tides, and saw the ocean running out of the Bay of Fundy, leaving whole deserts of mud flats exposed at low tide. I have mostly lived in temperate climates, where everything ices in winter, the world blooms in spring, the summers steam hot, and the leaves shower to the ground in the fall. All these changes in my environment—the sun, the moon, the seasons—are very vivid to me.

If you went to the moon to live, you would have to spend all your time "indoors" in an artificially created environment of a lunar station. But if you think about it, you will realize that many, maybe most, of us spend the majority of our time indoors in artificial environments. We warm ourselves with furnaces. We cool our filtered atmosphere with air conditioners. We light the night with fires and candles and fluorescent and incandescent lamps and bottles of fireflies. We zip around in jet planes. And a few of us escape the earth altogether to explore space. We are already living in space stations of our own creation.

We have not escaped time. Indeed, we closely manage our time. Clocks are important to us. When I was in England, I visited the oldest surviving clock in Salisbury Cathedral. It has no dial, no hands, and no case. The clock rings a bell to strike the hours and uses a toothed crown wheel and a regulating weight.[22] We carry watches. We keep and consult calendars. When I go to the beach in New Jersey, I consult tide tables so that I will know when to look for the tide pools at low tide.

When I was ordered to reduce my blood pressure by lowering the "stress" in my life, I realized that most of my own "stress" had to do with time. Do Bee on time. Don't Bee late. Cheese on bread! A *deadline* originally designated a line in a military prison which, if overstepped by a prisoner, would result in the prisoner being shot by a guard!

I had always been "on time." On time for class. On time to teach. On time for the movies. Heck, I even studied time!

We pay a lot of attention to time. I once heard this psychology joke: If you are early, you're anxious. If you are late, you are hostile. And if you are on time, you are obsessive compulsive. We end up "punching" a clock.

We are forever running, looking at our wristwatches, and crying frantically "I'm late, I'm late." Alice watched as the Rabbit pulled a watch out of its waistcoat pocket, consulted it, and scurried on. Louis Carroll's clock watchers may be rabbits with waistcoats, but rhythmickers have spent more time watching sea hares.

What is a Biological Clock

This book is not about mechanical watches and calendars. Even if you throw away your watches and calendars, time will play a part in your life. Animals and plants do not carry mechanical watches or consult printed calendars, but they are able to do things, such as feed, at certain times of the day, and they are able to do other things, such as reproduce, at certain times of the year.

The phrase "biological clock" has been used to denote these timing abilities.

Coleman defined a biological clock as "an innate physiological system capable of measuring the passage of time in a living organism."[23] What does that mean? "Innate" means the biological clock is inside the animal or plant. A "physiological system" means that the clockworks are anatomical parts of cells made up of bio-chemical machinery. Coleman's definition, however, measurement of the "passage of time," describes an *interval timer*, which simply measures the duration of a period of time, much as the sands run out in an hourglass.

The word *cycle* brings in the concept of some sequence of events, an increase and a decrease. One increase and decrease is one cycle. Thus, one night and one daytime make one daily cycle. One year, 365 days, make one annual cycle.

The word *rhythm* connotes recurrence. The rhythm of the tides, the rhythm of a waltz. A rhythm is constructed with repeated cycles. (I remember how to spell rhythm, because it has its own cycle of two "h" letters, as opposed to rhyme, used to describe poetry, which has only one "h").[24]

Both words, *rhythm* and *cycle*, have general usages which are not always biological. So we also talk about cycles of basic training or the rhythm of a poem, but they are not the subject here.

There is a more rigorous concept of a biological clock that has been developed by investigators which deals with cyclic events. In order for a biological clock to function, it must reckon time without being disrupted by environmental parameters, such as temperature, which are constantly changing.

The phrase **biological clock** can include clocks that measure days or years, but the concept was mainly developed for rhythms of about a day, which have been named **circadian rhythms**. Biological clocks have four features.

First, the central idea of a circadian biological clock is the notion that it is **endogenous**. An organism is capable of generating a rhythm (recurring cycles), with period lengths near 24 hours. The period lengths are determined innately by some mechanism within the organism. This viewpoint has been hotly debated. Proponents of the innate, or endogenous, biological clock have cited as their principal bit of evidence the fact that individuals generate rhythms with individually distinct period lengths. For example, one house sparrow might have a 23-hour rhythm, another house sparrow might have a 25-hour rhythm.

Second, a biological clock is nearly **temperature independent**. This is quite surprising since many aspects of living organisms are dependent upon chemical reactions, and chemical reaction rates usually depend on temperature. Generally, a chemical reaction proceeds faster at higher temperatures and slower at colder temperatures. The problem for a clock dependent on chemical reactions is obvious—it would run faster or slower as the temperature varied. Changes in the weather would play havoc, The clock's accuracy would be destroyed. Still, biologists found it difficult to give up the idea that the biological clock depended on temperature reactions, that there were some effects of temperature, so they coined the phrase *temperature compensated.*

Third, to be useful to an organism, it is reasonable that just as we must be able to reset our watches and clocks, so also must there be mechanism(s) for **resetting** biological clocks in order to synchronize them with events in the environment.

Fourth, the biological clock is viewed as a continuously consulted chronometer. That is, an organism can use its circadian biological clock to tell what time of day or night it is, much as we consult a watch or clock to find out the time. An example that led to this idea comes from honeybees, which possess a time memory, **Zeitgedächtnis**. When bees find a source of nectar, they can return to it at the same time of day at which it was originally discovered.

Why Ask Why?

Little children and college students always ask the hard question, "Why?" Why do we need these biological clocks? My honest answer is "I don't know" *why* they are there. What we know is that they *are* there.

We can speculate that biological clocks have evolved. Since the rotation of the earth occurs on a 24-hour schedule, many environmental events recur each 24 hours: light, dark, warm temperatures, cool temperatures, etc. These conditions affect organisms. For example, plants require light for photosynthesis and so they photosynthesize in the daytime when it is light.

It seems logical, that living things have organized their physiology on time bases (daily, seasonal) that correspond to cycles present in the environment. The ability to take advantage of the environmental cycles should have adaptive value. It should be useful for an organism to be able to predict the timing of the recurrence of environmental events. Anticipation would permit preparation. Therefore we can make an argument for why natural selection would favor the evolution of biological clocks capable of keeping track of and predicting the cycles that are present in the environment.

Bünning discussed an hypothesis of how the actual selection of clocks might have occurred. He suggested in the adaptation to the normal 24-hour period in the earth's environment, selection chose among a great variety of possible innate biochemical or biophysical oscillations. The rhythms already existed. It was not necessary to construct special hourglasses.[25] One means to separate, or compartmentalize, incompatible events in a physiological process, is to organize them with respect to time into sequences and oscillations.

A way to look at this is to imagine aliens from other planets. Other planets do not rotate with the same period as the earth. *Rotation* is the spinning of a planet. Mercury rotates every 58.6 days, Venus 127 days, Mars 24 hours 37 minutes and 22.6 seconds, and Jupiter 10 hours and 40 minutes.[26]

Earth's organisms have circadian rhythms, about a day in length. We would expect our imaginary aliens to have physiological adaptations to their own planets' rotation periods.

Similarly, the amount of time it takes a planet to orbit the Sun, its *revolution*, is variable. The earth takes 365 days. But Mercury zips around in 88 days, Venus speeds around in 225 days, Mars crawls for 687 days, and Pluto crawls for 90,780 days (248.5 years). Earth's organisms have **circannual rhythms** close to a year in length, we would expect our aliens from Mercury to have rhythms closer to 88 days.

Most of the biological rhythms I will discuss are adapted to period lengths characteristic of the earth.

Interestingly, life span, generation time, meal cycle length, sleep cycle time, gut beat time, breath time, and pulse time all increase as body or brain mass increases. Maybe there is an "evolutionary urge" to being active early because of intraspecific competition (the early bird, by being there first, gets the worm).[27]

Do we need our biological clocks to survive? We can speculate about the survival value of periodic functioning in organisms. There is an advantage to synchronizing a timed event among members of a population. That's the reason we carry watches, isn't it?, to make the synchronization even more precise. Temporal adjustments may save energy that would be otherwise wasted by pursuing the wrong activities at inauspicious times. But we really don't know how essential our biological clocks are for life, though we can, and do, speculate that they are valuable.

Chapter #3: Methods of Rhythm Measurement

In this chapter, I will discuss the way time is measured when we are talking about circadian rhythms.

Terminology and Abbreviations

Folks who study biological rhythms developed their own language, or jargon. The reader should not be discouraged by this lingo. The basics are simple.

Time can be thought of as a line extending backward (to the left) into the past, and forward (to the right) into the future. I think of time that way, like the print in this line, left to right.

A rhythm that persists in constant conditions is said to be *freerunning.* I love that beautiful anarchic word! For example, if you descended into a cave and were thus isolated from daily light and temperature changes and removed your all important watch, you will nevertheless still have a rhythm of sleeping and waking that would continue, or *freerun.* The Greek letter tau represents the length of the freerunning period. How often a cycle recurs, the number of cycles per day, is the reciprocal of tau, or frequency. In this book, for the most part, I talk about period length, and not about frequency.

When the timing of a freerunning rhythm is synchronized by a repeating external time signal, the rhythm is *entrained. Entrainment* is the process of being synchronized. You are most likely entrained in your normal daily life, probably by your use of your alarm clock. Plants and animals lack mechanical alarm clocks, but daily sunrise and sunset provide unambiguous precise time signals. The period length of the rhythm providing the entraining signal is designated by the letter T. In our normal environment, $T = 24.0$ hours, exactly.

A signal that gives time cues is called by a special name, the German word for timegiver, *Zeitgeber.* Light is the principal Zeitgeber for most daily rhythms.

A particular time point of a rhythm is a *phase*. The time difference between two phases is called the *phase angle*. When a rhythm is altered so that its peaks occur later in time, the phase of a rhythm is delayed; when a rhythm is altered so that its peaks occur early in time, the rhythm is advanced, a *phase advance*. The application of these ideas is clear in people. When we travel east across time zones (as when an American flies to Europe), we encounter earlier sunrises with respect to our time back home. Our internal rhythms must advance in order to synchronize with the new external time zone at our destination. When we journey westward (as when a European flied to America), a delay is required for our rhythms to synchronize, a *phase delay*. Advances and delays are called *phase shifts*.

There are abbreviations to represent the environmental lighting. LD means a cycle of alternating light and dark. LD 16:8 means 16 hours of light alternating with 8 hours of dark. DD means constant darkness. LL means constant light.

Clever Devices

The timing of human physiology may be of most interest to the reader (we are always interested in ourselves!), most of the experiments leading to our understanding of biological clocks has been with plants and animals.

In order to study a rhythm in some variable, investigators were required to measure that parameter as time progressed. A collection of measurements taken at intervals as time progresses is called a time series. In order to obtain a time series, investigators have sometimes taken a sampling of a population of organisms at different times. Another method has been to continually record a variable from an individual.

Sometimes the scientists met this temporal challenge the hard way, by working around the clock day after night after day. The difficulty of doing around the clock marathons inspired human ingenuity. The scientists invented automated sampling methods. Whew! These methods have permitted long-term recordings that last from weeks to years!

Orwell

The need for continuous monitoring has also predisposed the investigators to study rhythms (e.g. locomotor activity and body temperature) that were easily measured continuously over long periods of time. This section describes some methods for rhythm measurement.

An organisms' movement from place to place is called its *locomotor activity*. You have to watch out for the word "activity" because it is used more loosely and generally. Motions have been a popular parameter for study. An animal's habits are used.

Earthworms (*Lumbricus*) crawl 10 centimeters in a little over 21 seconds at midnight, but make the 10 centimeter dash in less than 17 seconds at 7 a.m. and the biggest day/night differential is in August.[28]

A perching bird's locomotion can easily be studied in the laboratory by recording from electrical switches attached to perches in the individual's bird cage.

Experimental regimens[29] are easily supplied with lights and timers.

Locomotor rhythms are measured in an individual rodent by attaching a recording switch to the wheel in a cage, or by making the entire cage a wheel. This method has not only been used to monitor the activity of rodents (rats, mice, hamsters)—you probably had a pet hamster with a wheel—but wheels have been used to measure rhythms of sea hares, lizards, and even cockroaches!

Alternatively, activity was also measured by placing a crab in a cage balanced on a "knife edge" so that its weight tilts the case against a switch each time the animal moves from one side to the other.

Plant leaf movements were measured by fastening a string from a leaf to a needle that scratched smoke from paper on a revolving drum. Other rhythms—oxygen consumption or carbon dioxide release—were measured by sampling the atmosphere of a plant.

Rhythms from populations of a marine dinoflagellate (a single celled alga) named *Gonyaulax* were measured another way. *Gonyaulax* are capable of bioluminescence, they glow at night. This rhythm can be monitored in a culture of *Gonyaulax* by recording the light with photocells.

Fruit flies have a rhythm of when they emerge, or hatch, from their pupal cases after metamorphosis, a process named *eclosion*. If pupae are stuck to tape, the emerging flies can be gathered at time intervals by shaking them into collection vials for later counting. The name "bang boxes" was given one such apparatus because the shaking involved banging on the boxes.

Some animals have vivid daily cycles in their body coloration because of melanophore cells in their skins. The common lizard sold in pet stores is *Anolis carolinensis*, the Florida chameleon. This lizard can change its skin color from green to brown in minutes. But it also has a daily cycle. The lizards are green at night (in the dark). In the daytime, the lizards' color depends on the background on which they rest. We examined lizard color rhythms by observing them periodically and with time lapse photography.

Human circadian rhythms have also been measured and recorded continuously. Often, the "volunteers" were the scientists themselves. Wever[30] described more than 200 "isolation" experiments in underground apartments. Environmental noise of less than 130 dB was excluded by constructing the units with double walls and separate floors, walls, and ceilings, so that each unit "floated" in a cocoon of glass wool. People spent 10–89 days in the units. Though they could leave anytime, only three of 232 "subjects" left because of the isolation. Because they were "volunteers" they did not represent the "average" population.

Investigators of human circadian rhythms often report the recording of "deep body temperature." When deep body temperature was recorded in house sparrows or rodents, implanted telemeters were used. However, for humans, a "rectal probe (resistance thermometer)" with a "sufficiently long wire connecting the probe with the recording system outside the experimental unit" was used. The locomotor activity was recorded with "contact plates on the floor invisible under carpets."

It has also been possible to record wrist motions from humans living their normal daily lives. The wrist monitors weighed three ounces, were metal black boxes fastened to the non-preferred wrist (in my case the left wrist) by a pair of Velcro bands. The wrist monitors were a conversation pieces. Most people who dared to wonder thought I was a convict on an electronic tether.

If you think about it (and I have) that means that electronic monitoring of criminals, and electronic surveillance (don't you think

Big Brother is watching?), have probably produced circadian rhythm data which are going to waste somewhere in electronic data banks.

Table 3.1 Examples of Some Persistent Circadian Rhythms.

Rhythm	Scientific Name	Common Name
Petals raised in daytime	*Kalanchoe blossfeldiana*	Plant
Leaf drooping at night	*Canavalia ensiformis*	Large beans
	Phaseolis coccineus	Beans
	Nicotiana tabacum	Tobacco
Pale body color at night	*Ligia baudiniana*	Isopods
	Uca	Fiddler crab
	Anolis carolinensis	Lizard
Intense running activity at night	*Mesocricetus auratus*	Hamster
	Glaucomys volans	Flying squirrel
	Periplaneta americana	Cockroach
	Carcinus maenas	Shore crab
Morning emergence from pupae	*Drosophila*	Fruit flies
Night CO_2 fixation	*Bryophyllum fedtschenkoi*	Succulent plant
Bioluminescence at night	*Gonyaulax polyedra*	Dinoflagellate
Daytime perching activity	*Passer domesticus*	House sparrow
Mating activity	*Paramecium aurelia*	Protozoan
Body temperature	*Myotis lucifigus*	Bat
Spore discharge	*Oedogoniuim cardiacum*	Alga
Phototaxis, peak in daytime	*Euglena gracilus*	Infusorian
Growth rate, peak in daytime	*Daucus carota*	Carrot
Volume of nuclei, peak in daytime	*Allium cepa*	Onion
Reduced night leaf heat resistance	*Kalanchoe blossfeldiana*	Plant
Night pineal melatonin synthesis	*Gallus domesticus*	Chicken
	Rattus norvegicus	Rat

Recording devices were used to make permanent records of the rhythms. Leaf rhythms were recorded with a smoked drum kymograph. Some rhythms were recorded with an analogue tracing, such as the bioluminescence rhythm. Event records, such as those made by Esterline-Angus®, were employed to record simple on-off events (e.g. wheel revolutions, perch-hops). The data were displayed using raster graphs. Computerized technology (e.g. Mini-Mitter Systems and Dataquest Software, Sunriver Oregon) accomplished the same task.

Cast of Characters

Humans, of course, have been studied. A plethora of human rhythms have been measured in a variety of experimental conditions, where the people were "isolated" from environmental time cues, and, less extensively, in people living freely on their own.

The wheel-running activity of laboratory rats and hamsters has been measured. Wheel-running activity of wild-caught rodents has also been studied, for example, white-footed mice (*Peromyscus leucopus*).[31] Some rat records (e.g. those of C. P. Richter) represent rats caught from the wild environs of Baltimore. Most rodents studied, indeed, most mammals studied extensively, have been *nocturnal* (night-active) animals.

Passer domesticus, the house sparrow, has been described as the major granivorous pest. Wild-caught sparrows are day-active, or *diurnal*, and have been a popular bird for study along with canaries and starlings. Their perch-hopping patterns are readily recorded.

Birds and mammals have a plethora of daily cycles in their behavior—aggression, coprophagy, copulation, courtship, drinking, eating, egg laying, food storage, gnawing, incubation, maternal behavior, nest building, nest departure, parental feeding, predation, preening, sex, singing, sleep, territorial defense, tonic immobility, torpor.[32]

Among the invertebrates, marine organisms (e.g. crabs) present special opportunities because they may be subject not only to 24 hour fluctuations in light and temperature, but to 12.4 hour fluctuations in the tides. In addition, many invertebrates have been attractive subjects because scientists had the notion that their physiology would be simpler, that their components, especially their nervous systems, would be less complex than those of vertebrates.

Marine mollusks, especially the sea hare (*Aplysia*) and its cousins, have gained attention for this reason.

A legion of plant species have been studied. *Acetabularia* is a single-celled plant whose one cell has been of particular use because its cell nucleus resided in one region that can be removed.

Neurospora, which is a bread mold, has served the science of genetics. It has rhythms of sporulation that can be observed in a "race tube." Growth medium is fixed to one side of a foot-long glass tube about 3/4 inch in diameter. *Neurospora* are added at one end. As the mold grows down the tube, it displays peach-colored bands of sporulation at daily intervals. If *Neurospora* is grown in a Petri dish, the daily sporulations appear as rings.

Lizards, reptiles, and fish, like the invertebrates, have body temperatures that depend upon the environmental temperature. For that reason, they have been used to study temperature as a Zeitgeber.

Graphs

Rhythms of organisms are represented graphically. Some of these graphs are shown in the section of figures. A rhythm can be represented with a longitudinal graph, as a sine wave, but the most useful representation, is as a raster graph (See Figures).

In a **raster** graph, time reads like a book, left to right, and top to bottom. The effect of this method is powerful. It permits instant assessment of entrainment and of phase shifts. By convention, the time of day is presented on the horizontal axis (*x*-axis, abscissa) and the days are arranged vertically in chronological order to that the vertical axis (*y*-axis, ordinate) is time in days. The convention is to place the most recent day's data at the bottom of the array. In addition, sometimes the records are duplicated horizontally to permit a rhythm to be observed without interruption.

Statistics

Don't be frightened! I have included this section, in case you are interested, because some readers may know about time series analyses or statistics from another field. This section is not essential for understanding this book.

Scientists being, well, scientists, it is not surprising that statistical analyses have been applied to rhythms. Not only have they made use of simple statistics—averages, calculation of variation, tests of significance—they have made use of specialized statistics developed for studying temporal events.

These specialized analyses are generally referred to as *time series analyses*. The rule of thumb is that it is desirable to have at least ten cycles of data for analyses.

The simplest analytical method that is appropriate, not for detection of rhythmicity, but to describe many cycles of circadian data as one cycle, has been to make an *average* curve, or average wave form. To make a graph of the average waveform, the period over which the average is to be made must be pre-selected by the investigator. Typical selections would be 24.0 hours for rhythms entrained to 24 hour regimens or the freerunning period (tau) for rhythms measured in constant light or dark. To make the average curve, an array of numbers is constructed and the numbers are summed and averaged vertically for each time point. The averaging technique is useful for discovering the shape of the rhythm. An entrained daily rhythm is not always sinusoidal. In house sparrows, for example, kept in LD 12:12, it is usually a bimodal pattern. There are two peaks, a big one after dawn, and a secondary peak of activity just before dusk. Most bird watchers know that good times to look for birds are just after dawn and as dusk approaches.

Seeking evermore objective means of assessing whether rhythms exist in biological data, James Enright was the champion for a technique for analyzing rhythm data.[33] In this method, arrays (or rasters) of data are made and average curves are calculated for a series of period lengths, for example, period lengths from 0.1 to 72.0 hours in 0.1 hour increments. The amount of variation present in each average curve is calculated. When the period length of the average curve matches the length of the rhythm in the data, the average curve has the greatest variation. A graph of the variation versus the periods is a *periodogram*. Peaks do occur at multiples and submultiples of the rhythm and there are subpeaks due to the waveform which must be taken into consideration when making an interpretation.

An *autocorrelation* method has also been used to analyze rhythms. In this method, the data are arrayed in a longitudinal time series. As with the periodogram, a series of period lengths are selected. For each period length, each data point is multiplied by the data point that period length distant in time. If the points are *maxima (peaks)* of a rhythm, then the product is large; if the points are *minima (nadirs)*, the product is small. For a given period length (called a *lag*), the products are summed and averaged. A plot of the resulting values is an *autocorrelogram*.

Fourier analyses have been applied to circadian rhythm data. To obtain a *power spectrum*, sinusoidal-shaped waves are generated with formulas so that the period lengths of the waves ranged from 0.1 hours to 72.0 hours (in 0.1 hour increments, lags again). Each of these waves is multiplied against previously calculated autocorrelation-wave. When the period of the formula-wave matches the period of the cycle autocorrelation-wave, the product of the multiplications is larger than when the period of the formula-wave and the autocorrelation-wave were not the same.[34]

The *cosinor* method consist of determining by least squares what cosine wave best matches a time series of data points.[35] The method can be used with very few points (e.g. samples at 4-hour intervals), even with unequally spaced data, and even with only one cycle of data, but the meaning derived from the results using sparse data is, of course, less reliable than when more data are used. The peak of the sine wave that best matches the examined data is called the *acrophase*. The acrophase is a powerful measure, because it permits comparison of peak times, from day to day.

Chapter #4: Entrainment

Under most circumstances, our daily rhythms, and those of the plants and animals that share our world, are synchronized, or entrained, by the daily rising and setting of the sun. Daily light provides precise time cues. The topic of *entrainment* is the one most relevant to your daily life.

When some people first discover data showing a freerunning biological rhythm, they jump to the conclusion that the freerun represents the natural situation, are attracted by that, and further aspire themselves to attain the freerunning state as somehow more natural and worthy. However, in nature, biological rhythms rarely freerun; synchronization is the rule.

The ability to freerun with a period (t) slightly different than the period of the entraining signal (Zeitgeber, T) may be a prerequisite for the ability to entrain, an essential feature of the clock.

Masquerade

There are also responses to cycles that do not involve synchronization of an innate biological clock. I have called the non-clock phenomena direct responses to light.

An example is that some sparrows hop on their perches when the lights are turned on, and they stop hopping on their perches when the lights are turned off. Most sparrows will hop during the light even if light is very short (e.g. three minutes of light presented every 10 minutes). However, following a cycle (such as LD 1.5:1.5), a sparrow placed in DD exhibits a circadian rhythm with a period near 24 hours, not a 3-hour rhythm. In other words, the sparrow has not "learned" a 3-hour rhythm. Direct control by dark also occurs in hamsters that run their wheels when the lights are turned off. You may think that this sounds pretty esoteric, that 3-hour cycles are pretty exotic, but consider that astronauts in a space shuttle orbiting the earth would see a 90-minute light dark cycle if they turned off all the artificial lights in their cabin.

When an external cycle, such as a light–dark cycle, synchronizes a biological rhythm, we use the word *entrained* to describe the synchronized biological rhythm. We say *entrainment* has taken place.

It is possible that both direct control and biological clock control occur simultaneously. It can be frustrating when it is difficult to decide which is occurring. The term *masking* has sometimes been applied to the situation—the biological rhythm is present but is occluded (masked) by the direct effect, or, the direct effect influences the measured parameter without altering the underlying biological rhythm process. To draw conclusions about what is controlling a rhythm, we would like to rule out the possibility of masking.

In considering a Zeitgeber, light in particular, we have to recognize that the signal itself has several parts—dark-to-light transition, lights-on, and light-to-dark transition. Pittendrigh[36] discriminates between two possibilities: (1) a continuous action of light, and (2) light being an effective signal only as it goes on and off. Both could in fact be operating.

Time Cues

A German word, Zeitgeber, which means "time-giver," denotes the environmental signals that provide time cues—lighting, temperature, food availability, sound, social factors, etc. In the natural world, most of the potential Zeitgebers should recur at 24 hours (light, high temperature, social activity) and should reinforce one another in entraining a circadian rhythm in those organisms that are sensitive to more than one Zeitgeber.

Light

The most well known and most studied of the potential signals for circadian rhythm synchronization is the daily light–dark cycle provided by the rotation of the earth. This provides a precise daily signal. Moreover, at most latitudes, the durations of light and dark fluctuate seasonally in a precise way. In the laboratory, the circadian rhythms of most organisms entrain readily to 24-hour light–dark cycles.

The photoreceptor used for this by mammals is the obvious one, the eyes. Preventing light of a light–dark cycle from being perceived by the eyes results in a freerun; for example, Richter's blinded rats freeran in light–dark cycles. Amazingly, some vertebrate organisms (among the birds, reptiles, amphibians, fish) synchronize to light–dark cycles even without their eyes, which means that there must be other photoreceptors, which has been called extraretinal light perception.

Temperature

As a result of the daily and seasonal cycles of sunlight, there are less precise daily and seasonal temperature cycles that provide potential Zeitgebers. It seems reasonable that some organisms should be able to use the cyclic changes in temperature as time cues. Some organisms, such as some lizards, can entrain to very low amplitude temperature cycles—less than a degree change per cycle.

Social Cues

Social factors have also been proposed as Zeitgebers. Two sparrows in adjacent, but separate cages, had circadian rhythms that freeran together for a time. Moreover the sparrows entrained to cycles of recorded song. Thus, sound can act as a Zeitgeber.[37] When pairs of sparrows were placed together, many of the pairs exhibited only one freerun, and there were instances of two rhythms in 40–50 percent of the pairs.[38] Pairs of hamsters separated by a wire barrier tended to freerun separately.[39] The result, that some paired animals can freerun independently, leads to the astonishing conclusion that it is possible for an organism to run on its own biological clock time in the face of potentially synchronizing time cues from its fellows.

Four humans kept for four days in constant dark with 24-hour sleep and feeding schedules stayed synchronized to 24 hours.[40] About half of blind folks, and even some people with vision, have 25 hour freerunning components in their activity/rest cycles.[41] Freerunning rhythms in a blind and sighted individual were found.[42]

Limits of Entrainment

A circadian rhythm can entrain to a 24-hour cycle. The length of the cycle providing the time cues is designated by the letter "T". It is also possible for circadian rhythms to entrain to cycles that are not exactly 24 hours in length. Experiments with exotic cycles of unnatural lengths are *T experiments*.

Circadian rhythms generally cannot entrain to all period lengths. There are limits of entrainment or a range of entrainment. Usually they can entrain to 22 or 20 hours (e.g. LD 11:11, $T = 22$ hours, or LD 10:10, $T = 20$ hours). But when cycles are imposed with T less than about 16 hours, the organism freeruns. The limits may be 21 or 26 hours in mice and hamsters, but in most other animals and plants, entrainment is possible down to 18 hours and the upper limit in plants, animals, and humans is about 28–30 hours.[43]

The effects of non-24-hour regimens can be dramatic. Nocturnal white mice advanced their onsets of activity and their maximal activity, compared to 24-hour regimens, when they were kept in 25-hour cycles. The normally dark-active mice became almost light-active. In a 22-hour day, the phases were delayed so that the mice were even more dark active.[44]

The exact limit is dependent upon the light intensity; it is possible to entrain to greater extremes of T with brighter light. You may think this is a pretty esoteric discussion. But our submariners use $T = 18$, an 18-hour day instead of a 24-hour day, for their cycles undersea. Most of our submariners are in three section watch rotations, three 6-hour watches.[45]

Frequency Demultiplication

Sometimes organisms kept in exotic short cycles (e.g. LD 6:6, LD 3:3, or LD 2:2) displayed a 24.0-hour cycle. This phenomenon is called frequency demultiplication. I have found more discussions of frequency demultiplication than examples. My own efforts to obtain an example were most fruitful driving the turtle heartbeat rhythm with electrical signals. The leaf rhythm of *Canavalia* persisted with a 24-hour rhythm in LD 6:6.[46]

Alarm Clocks

You are probably saying, "But I use an alarm clock!" Alarm use in university students ranged from 17–100% of days. On the days when alarms were used the average wake-up time was 7:12 a.m. EST which was 87 minutes earlier than the average wake-up time of 8:39 a.m. EST on days when alarms were not used.

Photoperiod

In a light–dark cycle, the length of the light-time, the day-length, is the *photoperiod* and the length of the dark-time, the night-length, is the *scotoperiod*. Entrainment is affected by photoperiod. For example, sparrows entrain to extremes of photoperiod, such as 1 hour of light, or 23 hours of light per 24 hours. When the photoperiod was six hours or more, the sparrow was active throughout the photoperiod. But in cycles of less than 6 hours of light per cycle, the sparrow exhibited activity in the dark which anticipated and trailed lights-on.

We are probably most aware of photoperiod when the days get short near the winter solstice, and when they get long near the summer solstice.

Phase Angles

The relationship between the entraining cycle and an entrained rhythm can be described and quantified by selecting a phase reference point in the entraining cycle (e.g. lights-on) and a phase reference point in the measured rhythm (e.g. onset of perch-hopping in a house sparrow). In sparrows, phase angle is a function of photoperiod—it is greater in LD 1:23 than in LD 4:20. The waveform was also affected. More sparrows were bimodal when the photoperiod was not extremely long or short.[47] This may seem irrelevant to you, but most people have a phase angle difference between the natural lights-on, that is, dawn, and the time they wake up.

Re-entrainment

When we travel east, the sun rises and sets earlier at our destination. In other words, the natural light cycle at the destination is advanced with respect to the cycle we were used to at home. Thus, in traveling eastward, we experience a short night followed by a new cycle.

When we travel west, the sun rises later than at home, and we experience a long day followed by a new cycle. In other words, we undergo a phase shift of the Zeitgeber(s) in our environment. We notice this most when the travel is across time zones and we have to reset our watches.

North–South travel within a time zone (New York to Florida) does not entail this phase-shifting phenomena, but the photoperiod changes because of the change in latitude.

Some journeys would result in changes of both phase and photoperiod. The internal circadian rhythms must reset to regain synchronization with the Zeitgebers in the environment.

We can mimic travel in the laboratory by changing the timing of entraining cycles. The organism adapts to the new cycle. Often there are intermediate cycles as the adjustment is made. For example, when a sparrow kept in dim LD 12:12 where the light intensity of L was 240 lux, intermediate cycles were seen for three days.

The word *transients* has been applied to these temporary oscillations or intermediate cycles. The entrainment before and after is sometimes called the *steady state*.

Circadian biologists have conventions for denoting phase shifts. They set 24 hours = 360 degrees and denote delaying phase shifts by minus and advancing phase shifts as plus. A 6-hour delay becomes a −90 degree phase shift. **Phase reversal** means a 180 degree phase shift, reversing the times of light and dark.

The rate or readaptation is a function of the conditions, such as the light intensity. For example, sparrows re-entrained to a 180 degree phase shift of LD 12:12 (240 lux L) in 1.4 ± 0.4 days if the light was extended to make the phase shift. But if the dark was extended for 24 hours to make the phase shift, the shift took much longer, 5.0 ± 0.8 days.

The interval of readaptation during travel is popularly referred to as **jet lag**. Travel, however, is not the only area of human endeavor where the rate of phase shifting is important. In order to provide 24-hour services or make 24-hour use of expensive equipment, many industries have made a practice of using various shift work schedules to obtain a continuous labor force. Often the schedules require frequent resynchronization of individual workers. Obviously, it is desirable to minimize the consequences of time shifting in travel or shift work.

Table 4.1 Rates of Phase Shifting.[48]

Organism	Days
Human	11
Rat	5–8
Mouse	9
Crab	6
Chicken	3–7
Sparrow	4
Roach	4

The number of days required for resynchronization after a 12 hour phase shift is shown. This is a phase reversal, or 180 degree phase shift. The rates of shifting vary not only upon the species, but upon the rhythm measured, and the nature of the Zeitgeber.

Skeletons

It is possible to mimic some effects of light–dark cycles with pulses of light. A skeleton photoperiod is created by using a pulse of light to represent dawn and another pulse of light to represent dusk. The amazing thing is that in animal experiments, skeletons work for both entrainment and photoperiodic control of reproduction.

Sometimes a skeleton is ambiguous. LD 1:6:1:16 could be the skeleton of either LD 8:16 (a short photoperiod) or of LD 18:6 (a long photoperiod). When a sparrow was presented with this choice it invariably chose the "short" photoperiod. This has been called the bistability phenomenon and a minimum tolerable night length has been proposed (about 10.3 hours for fruit flies).[49]

Incomplete Entrainment

Sometimes a signal is right at the threshold for entrainment—strong enough to catch a rhythm but too weak to hold it.

In *relative coordination* (beat phenomenon, oscillatory freerun) the organisms' rhythm remains roughly synchronized by the cycle, but the rhythm drifts a bit away from the signal and then periodically regains it.

Sometimes the organism freeruns past its Zeitgeber with a long period, then reverses direction to freerun with a short period to catch the signal. This is called *bounce*.[50]

The word *breakaway* has been used to denote the times when a rhythm freeruns in the presence of a signal to which it is sometimes synchronized.

Multiple Time Cues

In the natural environment there are cycles with a common period that provide reinforcing signals to an organism. For example, for the daily cycle, light and high temperature usually coincide.

An organism in the natural environment is subject to potential external Zeitgebers for more than one rhythm at a time. For example, a 24-hour cycle from the sun, a 12.4-hour cycle from the moon, and an annual cycle of photoperiod from the winter to the summer solstice.

There is more than one internal rhythm. For example, a woman experiences menstrual as well as daily cycles.

Thus there are interactions of rhythms and effects of competing signals (conflicting Zeitgebers) upon a rhythm.

Superposition occurs when an exogenous cycle forces its period in a physiological variable that also exhibits a circadian rhythm. For example *Euglena* showed a 2-hour rhythm of mobility superimposed on its circadian rhythm.[51]

Scientists have exposed fruit flies and cockroaches to simultaneous light and temperature cycles, varying the phase relationship of the light and temperature. They learned that the fruit fly eclosion rhythm would follow the low point of a temperature cycle during the light-time. Cockroach wheel running activity followed the high point of a temperature cycle during the dark-time. But the fruit fly rhythm could not follow the temperature cycle in the dark and the roach rhythm failed to follow it in the light. There was a 180 degree *phase jump* that occurred at a *critical phase angle*. In the *zone of forbidden phase* the temperature cycle could not drive the rhythms. Fruit flies and roaches have body temperatures that fluctuate with environmental temperature. In contrast, we are homeotherms, and maintain a more constant body temperature[52] more independent of the environment.

Rats are normally nocturnal and feed in the dark. When they were kept in LD 12:12 or LL and were only fed for 2 hours in the light, the rats ran their wheels in anticipation of the feeding time. So it appears that for rats, anyway, food is an effective Zeitgeber. Rats presented with simultaneous 24- and 25-hour cycles of food gave mixed results, synchronizing to one or both cycles, with freerunning activity between the two cycles.[53]

Chapter #5: Freeruns

If you can grasp freeruns, you will have a grip on the whole mess.[54] Yet, in a sense, you really don't need to know about them, because it is unlikely that your personal biological clock will ever be freerunning. So this chapter is irrelevant to your daily life, but is important to understanding how your biological watch works.

Still, I would wager money that you *could* freerun if we took away your wristwatch and clocks and placed you in the proper cave or chamber which isolated you from external time cues.

The key observation that is the basis of the field of biological rhythm is that rhythms observed in nature *continue* in the laboratory under experimental *constant conditions* (unchanging temperature, lights always off). This *persistence* is evidence for a *biological clock*.

Some dictionary definitions of the word *clock* differentiate a clock from a watch. A watch is a clock carried about on a person. Since individual organisms are themselves able to generate rhythms, the biological clock can be thought of as a *biological watch*.

A special word, *freerun* (which is not yet found in most dictionaries, and which some authors write as two words, free run) denotes those rhythms that were observed to persist in constant conditions. And most daily rhythms do persist in constant conditions.

Each individual has its own personal period length of a circadian rhythm, such as locomotor activity. They are only rarely exactly equal to 24.0 hours, but are usually longer or shorter than 24.0 hours. Investigators who find a rhythm in an organism with period length exactly 24.0 hours go back to the drawing board to check whether their constant conditions are really constant, and whether the organism could be getting a 24.0-hour time cue that it wasn't meant to receive.

Lighting AD LIB

We control the timing of our own light–dark cycle with artificial lighting. Some animals can control the timing of events by disappearing like the White Rabbit into holes, or by entering a cave.

Experiments where animals were permitted to choose their lighting were called *self selection* experiments. When sparrows were placed in cages and given access to a perch that controlled their overhead fluorescent light, some of them chose a circadian rhythm of light. They turned on the lights during their daily active period. This is basically what people do in isolation chambers and in caves. But the light that was self-selected by the sparrows affected their period length, as compared to the period length of their perching rhythm in constant dark (DD).[55]

Spontaneous Changes in Period Length

If your watch was not accurate to 24.0 hours, and if you wound it but did not reset it, your watch would "freerun" with a very accurate period.

Biological watches are not as accurate as most mechanical watches in maintaining period length. When circadian rhythms have been measured over long periods of time—months and years—unexplained changes in period were observed.

In sparrows, for example a maximum freerunning period occurred about 80 days after the birds were placed in DD, but the period lengths of mice freerunning in DD were still shortening when 300 days had elapsed.[56] These unexplained changes have been called "spontaneous" changes in period length.

Aftereffects of Pretreatment Cycle

Sometimes period length changes immediately following a change in environment conditions such as transfer from a light cycle into constant dark, or a light pulse. Collectively, period length changes immediately following a pretreatment have been called *aftereffects*. For example, sparrows placed into DD after LD 12:12 often have first a shorter than 24-hour period which gradually lengthens in a few weeks to a longer period length. This phenomenon is so common in sparrows that it has been given a name. It is called a "knee."

Aftereffects also follow the imposition of a non-24-hour regimen. Hamster period lengths in DD were larger after 25-hour LD cycles than after 23-hour LD cycles.

Photoperiod has aftereffects. After 12–15-hour photoperiods, sparrows had shorter freeruns than after 2–6-hour photoperiods. After LD 16:8 the period length was 23.7 hours; after LD 8:16 the period length was 24.1 hours.

Activity Time and Rest Time

The length of a cycle of a freerun is called its *period length* (tau). The length of time in a cycle that the organism is active, its activity time, is called **alpha**. The length of time the organisms is inactive, its rest time, is called **rho**. Alpha plus rho equals the period length. *Alpha* for humans is essentially the duration of the awake time. *Rho* for humans is the duration of the sleep time.

It may surprise you to learn that the durations of these times are subject to environmental control. Moreover, the effects of the environment are *opposite* for diurnal organisms and nocturnal organisms.

In a light–dark cycle, a nocturnal animal, such as a hamster or mouse, is usually active (runs its wheel, eats, mates, etc.) in the dark. When the animals are placed in constant dark (DD), the activity time is usually longer than it is for hamsters placed in constant light.

In contrast, when a diurnal animal, such as the house sparrow, is placed in constant light (LL), its activity lasts longer (e.g. activity time, alpha = 17 hours) and its rest period is reduced in comparison to house sparrows in DD (e.g. activity time, alpha = 8 hours). Thus a *nine*-hour difference in activity time can be produced by light!

The duration of activity in constant light (LL) depends on the **light intensity**. In an experiment with chaffinches (*Fringilla coelebs L.*), the activity duration was 13.7 hours in dim light (1 lux) and the activity duration was 17.8 hours in bright light (5 lux).[57]

Sometimes it is useful to express the relationship of the duration of activity to the rest time as a ratio, the **alpha/rho** ratio. In the chaffinch experiment, the alpha/rho ratio in the dim light was $+1.4$, and in the bright light was -3.8.

Rule for Period Length

Period length is also dramatically changed by the quality of lighting. Here again, nocturnal and diurnal species are usually affected in opposite ways.

For example, diurnal sparrows usually have longer freerunning period lengths in constant dark (DD) and shorter freerunning period lengths in constant light (LL). Nocturnal hamsters have shorter freerunning period lengths in constant dark (DD) and longer freerunning period lengths in constant light (LL).

Just as for the alpha/rho ratio, the intensity of the light is important. The chaffinches had a period length of 23.4 hours in dimmer 1 lux LL and a shorter period of 22.5 hours in brighter 5 lux LL.[58]

Circadian Time

Circadian rhythm biologists have adopted a convention that is sometimes used to denote time of day on the internal biological watch. This is not the same as reckoning time according to the sun or the clock on the wall. It is useful for designating time in freerunning rhythms.

To determine circadian time, the scientist measures a freerunning rhythm. He then sets the period length of the rhythm (which is not 24.0 hours by the clock on the wall) equal to 24.0 units of circadian time. By further convention, circadian time zero for diurnal animals is set at activity onset. Circadian time 12 is designated as activity onset for nocturnal animals.

Chapter #6: More Related to Freeruns

The topics in this chapter—development, aging, accuracy, the clock and hands problem, two views, inherited clocks—relate to the concepts of freerunning circadian rhythms.

Development

When do circadian rhythms first appear during the development and maturation process?

Parents of human newborns are only too well aware of the irregular and disturbing sleep patterns of their offspring during the initial weeks. When sleeping patterns of an infant were recorded for the first month of life, the circadian rhythm was not established at birth. The rhythm freeran. It had a period of about 25 hours by six weeks of age, and it gradually became shorter and more synchronized with a 24-hour schedule.[59]

Freshly hatched chicks, *Gallus domesticus*, are precocious in that they can readily survive without a mother hen if kept in a warm environment and given access to chick feed and water. Daily cycles and circadian rhythms, for example of the pineal enzyme, N-acetyltransferase,[60] can be measured in chicks on their *first exposure to a light–dark* cycle. Indeed, when chicken eggs were kept in a light–dark cycle during incubation, it was possible to detect a cycle in the pineal glands of the embryos at 17 days of incubation, well before hatching.

The offspring of many mammals are not as self-sufficient at birth as chicks. The pineal N-acetyltransferase enzyme rhythm appears at around 4 days of age in rat pups. Movement rhythms were not present in rat pups during the first 10 days without their mothers, but they have rhythms earlier when kept with the rat dams and they are synchronized by rat mothers.

Aging

What is the effect of aging on circadian rhythms?

The amplitude of many rhythms decreases as old age approaches. Usually this decrease is a reduction of the maxima for the rhythm, for example, hamster pineal melatonin is nearly 1000 units at night in 2-month old hamsters, but less than 200 units in 18-month old hamsters.[61] Hamsters live about two years. Other rhythms that lose amplitude include: mouse body temperature, rat body temperature, mouse audiogenic convulsions, mouse oxygen consumption, human potassium excretion, human growth hormone, human testosterone, and human luteinizing hormone.[62]

Period length of the freerunning rhythm also changes with age. Rodent freerunning periods decreased from puberty to old age.[63]

The duration of rest time (sleep duration) decreases in humans over the course of their lives. Soon after birth, infants may sleep over sixteen hours a day, and by 12 months they may sleep 14 hours a day. The sleep duration drops to about 10 hours a day during the teen years and to a little over 7 hours a day by age 20.[64]

Accuracy

Enright considered the precision of the circadian biological clock and reported that three canaries, for example, had individual cycle-to-cycle accuracies with standard deviations of 3, 4, and 7 minutes. He described this as an error of as much as one part in 200 or as little as one part in 500. The precision of circadian rhythms was compared to that of other rhythms: the spontaneous firing of single neurons (more variable than one part in 50), the human heartbeat (more variable than one part in 60), the chirping of crickets (one part in 30), flashing of some fireflies (one part in 200), and the discharge pattern of electric fish (one part in 8000). The vertebrate cycle-to-cycle variability in period is often less than 1% of the period and predictable to within 2 to 3 minutes.[65]

Flying squirrels have individual circadian rhythm cycle-to-cycle variation as small as 2 minutes and as large as 85 minutes.[66] Cycle-to-cycle variation measured for 10 circadian cycles in house sparrows freerunning in DD was as small as 10 minutes or as large as 85 minutes.

Table 6.1   Circadian Periods for Ten Successive Cycles in a Sparrow Illustrate their Accuracy.

Cycle	Period Length (hours)	Difference (minutes)
1	24.50	30
2	23.84	− 10
3	25.02	61
4	24.32	19
5	24.65	39
6	24.43	26
7	24.41	24
8	24.96	57
9	24.27	16
10	25.42	85
Mean	24.68	37
Standard deviation	0.70	24
Standard error of the mean	0.22	8

Clock and Hands Problem

The observed rhythm, such as the perch hopping or wheel running activity, that we can measure, is like looking at the hands of a clock, not at its clockworks, or underlying mechanism. An indicator process is measured, and we can only make inferences about the controlling process from the indicator measurements.

Ubiquity

Where circadian rhythms have been sought, they have been found. Thus there is a long. list of organisms in which circadian rhythms have been studied. I counted over 150 species in one book and almost a hundred in another. There are also long lists of circadian rhythmic functions within individuals of a species. Circadian biologists like to say that circadian rhythms are "ubiquitous," they are everywhere.

Two Views

Early in the history of the field of circadian rhythms, two alternative hypotheses were put forward for the mechanisms responsible for the freerunning rhythms.[67]

The *exogenous hypothesis* was championed by Brown. Fiddler crabs (*Uca pugnax*) are dark colored in the day but turn pale at night so they respond to a daily light–dark cycle. But, remarkably, captive crabs kept in the laboratory had two periods of locomotor activity each day, about 12.4 hours apart, that followed the timing of the tides on the crabs' home beach even when the crabs were kept in a 24-hour light–dark cycle. Thus, it appeared that the crabs were able to synchronize with something in the environment that had the same frequencies as the tides.

An exogenous interpretation of freerun is that it originates from periodic events that do not recur at exactly 24 hours in the external environment. The idea is that freerunning organisms are synchronizing to these cues much as the crabs follow the timing of the tides. Collectively, these potential time cues were called *subtle geophysical factors*. The subtle geophysical factors include gravity, geomagnetism, atmospheric tides, electrostatic fields, and background radiation. There is evidence that organisms can detect such things as the local geomagnetic field. Rhythms whose freerunning periods are 24.8 hours are particular suspect because that is the period length of the lunar day.

In the endogenous hypothesis, the individual organisms are viewed as in possession of an internal cellular-biochemical oscillator that is responsible for freerunning circadian rhythms. They have an internal circadian pacemaker, much as the heart has a pacemaking sinoatrial node.

Compelling evidence for the endogenous view is that individual organisms freerun with their own personal period lengths (as for the sparrows). The role of the environmental cycles, in the endogenous view, is to provide time cues that can synchronize, or reset, the internal clocks.

Inherited Clocks

Genetic studies support the idea that circadian rhythms are inherited. Although there are some aftereffects of timed pretreatments, new frequencies are not learned.

Sixteen-hour (rather than 24-hour) cycles did not cause rats or plants to lose their ability to produce a subsequent daily rhythm. Successive generations of mice raised in constant light (LL) did not lose their circadian freeruns. Second generation mice had average period lengths of 25.5, 26.1, and 25.2 hours. Even rats kept in constant light (LL) for 25 generations did not lose their circadian rhythms.[68] Raising mice born in a 20-hour or 28-hour cycle did not affect their freerunning periods, activity times, and rest times.[69]

Hybrids were also studied. In some cases, intermediate periods were obtained in hybrids, suggesting that many genes participate. However, fruit fly mutants created with ethyl methane sulfonate had altered X-chromosomes. Some were arrhythmic, some had 19-hour rhythms, and others had 28-hour periods. The effects were attributed to a single gene on the X-chromosome. An amazingly similar **single gene locus**, called the *per* locus, was also suggested for *Neurospora* mutants, and a one gene hypothesis was also championed for *Chlamydomonas*.[70]

Mice were mutated with a chemical (ENU, N-ethyl-N-nitrosourea) producing a male that had a period an hour longer. Using him, they were able to breed homozygous mice with 27–28-hour period lengths which lost their rhythms after two weeks of constant dark. They called this mutation *Clock*.[71]

A single male Golden hamster (*Mesocricetus auratus*) was found with a 22-hour period length in constant dark. Normally, hamsters have period lengths no shorter than 23.5 hours (the average is 24.1 hours). Using breeding experiments beginning with this one king hamster, it was suggested that a mutation had occurred at a single location and this was named tau.[72]

Chapter # 7: Resetting

Resetting is basic to the functioning of the biological clock. At most latitudes, organisms face a change in Zeitgeber timing with each new day's dawning because the photoperiod has changed. Because an organism's freerunning period is not 24 hours, the organism must reset every day or at frequent intervals if it is to stay in synchrony with the natural day–night cycle of the environment. This discussion of resetting may seem a bit obscure, but resetting is pertinent to the two important challenges to human biological watches: jet lag and shift work.

Perturbing Pulses

A tool that has been used to study resetting has been the measurement of phase shifts following single short pulses (perturbations) of a Zeitgeber. The bulk of these experiments have used short light pulses (e.g. 10 minutes to 6 hours), but some investigators have employed dark pulses, temperature pulses, steps from light-to-dark, etc. Just as when the phase of an entraining cycle is shifted, following a pulse, transient cycles are sometimes seen before a new steady state is attained.

Phase Response Curves

When scientists methodically exposed organisms to single perturbing signals (e.g. light pulses) at different times over 24 hours, the phase shifts they obtained were not all the same. They graphed the direction and amount of phase shift obtained versus the time that the pulse was given. Such a graph is called a *phase response curve* or PRC.

Phase Response Curves, Light Pulses

When phase response curves were measured for light pulses in different species, common features became apparent. Delay phase shifts were found in the early half of the projected night. The time during which delays were found bracketed the time of expected dusk. Advance phase shifts were found in the late projected night, bracketing the time of expected dawn. There are times of apparent relative insensitivity to light around midnight (*singularity*) and midday (*dead zone*).

Phase response curves may represent the sensitivity of the underlying circadian pacemaking process to the Zeitgeber. However, it is alternatively possible that the variation in response revealed in a PRC represents a rhythm of sensitivity in detection of the Zeitgeber (e.g. the photoreceptor sensitivity in the case of light pulses).

You may think that PRC's and phase shifts are esoteric phenomena in the laboratory, but consider that you get a "light pulse" if you get up and turn on the lights in the middle of the night.

Phase Response Curves, Dark Pulses

While most phase response curves were made using light pulses, it was also possible to measure phase response curves using dark pulses. When that was done, it was found that the "sign" of the graphs was reversed, that is that advances were in the early subjective night, and the delays were in the late subjective night. Although the graphs are not exactly reversed, they have been referred to as *mirror images*.

Again, the effects of dark pulses may be relevant to you if you take a nap in the dark during the daytime.

Ways of Obtaining Phase Response Curves

Scientists have obtained phase response curves (PRCs) for circadian rhythms using a variety of techniques.

I want to make a PRC, let me count the ways:

(i) Application of a light pulse to the organisms freerunning in constant dark (DD)—this method has been popular with investigators studying hamsters and mice.

(ii) Application of light pulses to organisms in constant dark during the first few days of constant dark (DD) following pretreatment with a light–dark cycle (LD 12:12). This method was used by investigators studying fruit flies and sparrows.

(iii) Application of a light-step (a dark-to-light transition, L/D transition, or lights-on).

(iv) Calculation from a freerunning rhythm as it is perturbed by a daily weak signal.

(v) Calculation of the shift resulting from a dark-step (L/D transition, lights-off).

(vi) Application of a light pulse in the dark time of LD.

(vii) Exposure of the organisms to dark pulses in constant light.

When the organism is a hamster, mouse, or sparrow, the first locomotor activity onset in constant dark or the extrapolated line through onsets is used as the measurement from the animal's locomotor rhythm to determine the animal's actual phase.

The *phase shift* is measured, for example, in sparrows, as the time difference between the last onset or average of onsets before the signal, and the onset after the signal.

For rhythms, such as the temperature rhythm, for which there is no onset, a criterion must be set up to measure phase. The phrase *phase reference point* is used to denote the characteristic of the rhythm that is used to measure phase (e.g. onset of activity in hamsters or sparrows, or median of the eclosion peak in fruit flies).

Factors that Modify Phase Response Curves

Not surprisingly, PRC's are not all the same and scientists studied the factors that modified them.

PRCs have individual species characteristics in the magnitude of the phase shifts that are obtained in response to similar signals.

Nocturnal or diurnal life style is important. Larger phase shifts were measured with smaller signals in nocturnal animals. Nocturnal hamsters and flying squirrels respond to 15-minute light pulse. Larger signals, 2–4 hours, were required to get measurable phase shifts in diurnal house sparrows.

PRCs for nocturnal species should have a larger delaying portion in their PRCs and diurnal species should have a larger advancing portion to their PRCs. That is because the nocturnal animals generally have freeruns shorter than 24 hours and need to delay to correct to the 24.0-hour day, whereas diurnal animals generally have longer than 24.0-hour freeruns and need to advance to correct to the 24.0-hour day. For example, a sparrow with a 25-hour freerun need a one-hour advance each day in order to stay entrained.[73]

The quality of the signal changes the appearance of the PRCs. For example, fruit fly PRCs were measured using pulses with 12 hours, 4 hours, and 1/2000 second (!) durations. The longer signals produced larger phase shifts, and thus a PRC with a larger amplitude.

The photoperiod used to pretreat the organisms also determines the exact shape of the PRC. In sparrows, for example, the PRC obtained after long photoperiod (LD 16:8) has a large advance portion. The PRC obtained after short photoperiod (LD 8:16) has a substantial delay portion in the early projected night and the advancing portion is reduced. Hamsters pretreated with long photoperiod (LD 18:6) had a PRC with only 31 per cent of the amplitude of the PRC of hamsters pretreated with short photoperiod (LD 10:14).

Instantaneous Resetting

Temporary, intermediate, or *transient* cycles that are seen in the measured variable (e.g. the sparrow perch hopping rhythm) during a phase shift are thought to represent the "motion" of a "slave" or "driven" oscillation as it gradually regains phase with an underlying internal pacemaker oscillation, the biological clock.

The implication in this explanation of transient cycles, is that the phase shift in the pacemaker is *instantaneous*. That is, that the phase shift of the pacemaker is completed by the time the perturbation is completed.

Thus one can design experiments involving two pulses, a first pulse, and a second tester pulse. Such *two pulse experiments* have supported the hypothesis that the phase shifts of the pacemaker oscillation are completed immediately.[74]

Explanations Based on PRCs

One of the reasons that phase response curves capture the interests of circadian biologists is that they offered elegant explanations of fundamental observations of the properties of circadian rhythms, such as the effects of constant conditions on period length and the mechanism of entrainment.

Phase response curves were used to explain the effects of light intensity on period length and the alpha/rho ratio. Nocturnal animals (flying squirrels, hamsters, mice) have larger delaying portions of their PRCs, so in constant light (LL) the effect should be delaying and a longer period length. Diurnal sparrows have a larger advancing portion in their PRCs, so the effect of constant light (LL) should be advancing, producing a shorter period length in constant light (LL).

Various theories of entrainment have been based on the PRCs. For example, entrainment when the days are shortening can be explained by invoking the phase shifting sensitivity to dark revealed by the dark pulse PRC, and entrainment when the days are lengthening can be attributed to use of the sensitivity to light as mapped by the light pulse PRCs.

PRCs have also provided satisfying explanations for the curious effects of skeleton photoperiods and the bistability zone. Recall the example for sparrows in skeleton cycles. Sometimes a skeleton is ambiguous. LD 1:6:1:16 could be the skeleton of either LD 8:16 (a short photoperiod) or of LD 18:6 (a long photoperiod). When a sparrow was presented with this choice it invariable chose the "short" photoperiod. For the sparrow to choose the 18-hour photoperiod would mean that it would experience a light pulse falling about 7 hours into its subjective dark-time, which, from its PRC, would produce a big advance, and an unstable situation. The alternate choice of an 8-hour photoperiod produces stable entrainment.

Chapter # 8: Can the Clock be Stopped?

Do you have enough time?

One of the reasons that I found circadian rhythms has to do with what sparrows do in constant light. In constant light sparrows are constantly active. They lose their rhythms. I wondered, were their clocks stopped? Turtle hearts continue to beat without the rest of the turtle. Placed in the refrigerator, the heart first slows down then stops beating as it chills. If the heart is removed from the refrigerator the next day, when it rewarms it commences its beating. Clearly, the turtle heart rhythm can be "stopped" and "restarted." But what about the circadian clock? Can it be stopped?

In previous chapters, alterations of the freerunning period (slowing and speeding up the circadian biological clock) were discussed, as well as resetting reactions to disruptions, or perturbations, of the entraining signal (Zeitgeber). In this chapter, some effects of constant light will be discussed—apparent stopping of the clock, arrhythmia, and splitting of rhythms. Already mentioned, is the effect of constant dim light on period length: shortening it in diurnal house sparrows, lengthening period length in nocturnal hamsters. This chapter is probably irrelevant to your daily life, but contains phenomena that I found most interesting in all of circadiana.

Again, the phenomena discussed in this chapter may seem esoteric and far removed from your daily experience. But they have implications for the resetting of your clock and its running.

Arrhythmia

I was totally impressed by the response of house sparrows to constant light. In bright constant light (LL, e.g. 1000 lux, approximately your typical office fluorescent light), the perching activity of house sparrows becomes arrhythmic. The sparrows hop with gusto all 24 hours of the day and night, day after day, month after month. The sparrows didn't drop dead from this apparent lack of rest. They looked great. They all did it. There I was, sleeping some seven hours a night, having only two-thirds of the 24 hours at my disposal, and here were these sparrows with the full 24 hours to use.

The sparrows that were active through all 24 hours in constant bright light were, to all appearances, and time series analysis, arrhythmic.

The ability of sparrows to be active throughout 24 hours for weeks and months raises the question as to whether it would be possible for human beings to increase our usable daily time and reduce sleep need by exposing ourselves to constant bright light. My home experiment on myself ended prematurely after four days because of howling protests from my significant other who didn't like sleeping with the lights on. We may have already answered the question, to some extent, because by use of artificial lamplight in the evening, we humans have an average waking time of over sixteen hours a day even during the winter months when daylength is short.

What is the meaning of arrhythmia in LL for sparrows? Does it mean that the clock is stopped? Or is the clock running, but its expression is masked by a direct effect of light? I cannot answer that question now, and we may never be able to separate direct and circadian effects of light on sparrows.

You should not jump to the conclusion that living in constant light, or extended light, is desirable for either us or for the animals and plants that are exposed to our artificial lights. Especially for plants, damage due to LL has been observed—failure of flowers to open in the evening primrose, fungi unable to form sporangia, loss of chlorophyll in an alga. Some of these deleterious effects of LL on plants can be prevented by exposing the plants to a daily temperature cycle white they are in LL.

Often, upon placing an organisms in constant light (LL), the circadian rhythm is not lost immediately. Instead, the rhythm damps out more or less gradually over a few cycles, a so-called fade-out. It is also not always the case that LL produces arrhythmia, even when the LL produces damping of the amplitude of the oscillation.[75]

Splitting

When hamsters (or tree shrews or rats) are placed in constant light, their period lengthens. Their activity time, alpha, shortens, or is compressed. This makes their freerunning record look like a tornado. When male hamsters were kept in constant light for 30–100 days), the circadian rhythms split into two freerunning components with periods near 12 hours in 56% of them. Splitting was reversed within one to four days when the hamsters were returned to DD; the activity rhythm re-fused into a single component, usually with a longer period length than for the dissociated rhythms. The scientist who have studied splitting have attributed it to the presence of two oscillators, a morning (M) oscillator, and evening (E) oscillator.[76] Split rhythms echo the idea of a tidal component, and of bimodal patterns. But the bimodal patterns, as of house sparrows, have activity in the interval between two peaks, which is not seen in the splitters.

Resetting

When trying to explain arrhythmia, it might be impossible to separate the two possibilities, is the clock stopped? or, is the clock running, but the locomotor activity is arrhythmic because of a direct effect of light (masking).

We can transfer organisms from constant light (LL) to constant dark (DD), examine the freeruns, and determine the phase of the onset of the freerunning activity in DD.

After the lights-out of LL, sparrows first were inactive, as they are during a normal night. About 12 hours later, some of the sparrows commenced a circadian rhythm in perch-hopping activity, which continued to freerun in DD. The average duration of the rest period before the first perching after LL was 15 hours. The data are evidence that the circadian clock in sparrows stopped at the end of the first' subjective day, the time the birds expected dusk to occur (provocatively close to our sixteen-hour day).

However, there is another explanation, to wit, the biological clocks of arrhythmic sparrows are present, and at different individual phases, but are masked, and the L/D transition resets the individual sparrows by different amounts so that, after L/D, they all appear to have the same time setting.

So, the experiment adds information that has practical value— a way to quickly set sparrow clocks is with constant light—but it doesn't rule out the possibility of masking.

Sparrows placed in constant dark (DD) after LD 1.5:1.5 (in which the sparrows are active during the 1.5-hour light periods) produced results similar to the pretreatment with LL. The sparrows clock time zero was 15–16 hours after the last L/D.

Fruit flies are also reset by the L/D transition at the end of LL. The effect is so reliable and routine that it is used to synchronize the rhythms of individuals within populations of fruit flies. How long does the constant light have to be? Using any final photoperiod longer than 12 hours, the flies' eclosion rhythm was timed by the L/D. The new phase was always 15 hours after L/D. So for flies, 12 hours of light was sufficient for L/D to completely set the clock. The L/D transition was an absolute phase-giver.[77] Moreover, when the light used was shorter, the fruit flies were reset by the dusk (L/D, lights-off) in the flies' early night, and reset by dawn (D/L, lights-on transition) during the flies' late night.[78]

Singularity

Fruit fly resetting has been studied in great detail, methodically varying the signal duration of the pulse from 0–120 seconds. When this was done, it was found that there was a singularity. The singularity occurred at a critical stimulus time (called T*) at 6.8 hours (after L/D, that is about 7 hours into the night) with a duration of stimulus, S*, of 50 seconds. This is the time when the phase shift change from delays to advances. What was amazing, was that after this critical signal, T*S*, the phases of the flies' rhythms were unpredictable. When the flies emerged, the eclosion rhythm of the population of the flies was lost or annihilated. Presumably, the flies responded individually and were scattered to different phases because of the individual characteristics of each fly's biological clock. Or, perhaps, the pulse puts each fly into a phaseless state where it remains until another pulse reinitiates the oscillation and it can assume any phase.

Dissociation

The human body temperature rhythm was observed to freerun with a separate period length from the freerunning activity rhythm. The body temperature rhythm was dissociated from the sleep–wake rhythm. Dissociation produces freeruns with different period lengths; splitting produces two components which have the same period length.[79]

The explanation of dissociation is that the organism, in this case the human, contains more than one circadian oscillator or pacemaker.[80] Normally, these rhythms have the same period derived either from one another or from a common pacemaker. The word coupling refers to the physiological relationship by which two oscillations within an individual are synchronized. So, in dissociation, the temperature and activity rhythms were uncoupled.

Dissociation has also been claimed and used as an explanation for changes which occur during phase shifting, as the changes observed in jet lag. The idea is that after phase shifting by crossing time zones, the separate rhythms resynchronize at different rates depending upon the strength of their coupling to the pacemaking rhythm—stronger coupling means faster synchronization.

For humans, this can be serious because when rhythms are disrupted by travel or shift work, a whole raft of symptoms may be observed: fatigue, accidents, headaches, burning eyes, gastrointestinal problems, shortness of breath, sweating, nightmares, insomnia, menstrual irregularities. The idea is that these symptoms are a result of asynchrony among the individual's normal internal physiological rhythms.

Chapter #9: Temperature and Chemicals

Light is not the only thing that can control circadian and other rhythms, though it is probably the Zeitgeber that has been studied the most. This chapter explores some other things that can affect rhythms.

Temperature

The effects of temperature has been a subject of interest for investigators of hibernation, cryobiology, thermoregulation and for cell biologists and biochemists.

The biochemical machinery of organisms is dependent on chemical reactions. Heating increases the rate of a reaction. That fact provides a rationale for the evolution of organisms' ability to generate their own heat and to maintain their body temperatures above the temperatures in their environment.

Scientists have argued that biological reactions would still be too slow at normal temperatures for cells to function, so there are additional strategies for increasing reaction rates, such as concentrating the reactants, and the use of catalysts, called enzymes.

Typically, the activity of an enzyme is negligible near freezing, but increases with increasing temperature. Scientists quantitate the dependency of reaction rate on temperature with a value, the Q_{10}, that represents the effect on the rate of a reaction of a 10°C increase. If the reaction rate doubles, $Q_{10} = 2$. If there is no effect of temperature on a reaction, $Q_{10} = 1$.

Human body temperature is 37°C. At 40–60°C, the enzymes break down, they are inactivated, which is called denaturation. We routinely use cold to arrest the rate of biological reactions when we put food in the refrigerator or freezer. In the laboratory, the activity of an enzyme, such as the pineal enzyme, N-acetyltransferase, can be preserved by freezing pineal glands or by maintaining homogenates of pineal glands on ice.

What are some of the effects of temperature on circadian rhythms? There are some disparate aspects of temperature that particularly pertain to circadian rhythms.

Extreme heat or cold abolishes rhythmicity. For example, the rhythm of dinoflagellate bioluminescence in a culture of dinoflagellates was lost in the cold at 11.5°C. The amplitude of the bioluminescence rhythm was damped, or lessened, at 23.6°C and nearly abolished at 32°C. The largest amplitude was at about 32°C.[81]

However, there is surprising little effect of temperature upon period length. Q_{10} for circadian rhythm period length is usually near 1.

Table 9.1 Period Length is Only Slightly Altered by Temperature.[82,83]

Rhythm	Species	Temperature (°C)	Period Length (hours)
Cockroach running activity	*Periplaneta*	19	24.4
		29	25.8
Dinoflagellate bioluminescence	*Gonyaulax*	16	22.5
		22	25.3
		32	25.5
Bean leaf movements	*Phaseolus*	15	28.3
		20	28.0
		25	28.0
Lizard activity	*Lacerta*	16	25.2
		25	24.3
		35	24.2
Bread mold growth	*Neurospora*	24	22.0
		31	21.7

Table 9.2 Q_{10} Values for Circadian Period Length.

Q_{10}	Organism	Rhythm
0.8	Alga	Sporulation
1.0	Bats	Activity
1.0–1.3	Beans	Leaf movement
1.0	Bees	Feeding time sense
1.0	Bread mold	Growth zonation
1.0	Ciliate	Mating
1.0	Crayfish	Eye-pigment migration
1.0	Fiddler crab	Color change
1.0–1.1	Flagellate	Phototaxis
1.0–1.2	Fruit flies	Eclosion
3.0	Grasshopper	Hatching
1.1	Hamster	Activity
0.9	Marine dinoflagellate	Bioluminescence
0.9	Marine dinoflagellate	Cell division
1.1–1.4	Mouse	Activity
1.3–1.5	Mold	Sporulation
1.0	Oats	Growth rate
1.0	Roaches	Activity
1.1	Sunflowers	Exudation

At first this evidence appears to indicate that the circadian period length is *independent* of temperature. It is not dependent upon a chemical reaction, rather, the Q_{10} implicates a physical process such as diffusion.

However, because the Q_{10} is not exactly 1, rhythmologists have argued for a process of temperature *compensation* in which two or more opposing processes are involved so that the net effect is very little effect of temperature.

It makes sense for circadian period length to be relatively independent of environmental temperature, because it would reduce accuracy if the speed of the biological clock were subject to the gypsy whims of Mother Nature's weather.

Organisms, especially those that are not warm-blooded, such as the invertebrates, can be cooled to 0–5°C, just above freezing. Cold temperatures retarded (delayed) the clocks (of spiders, roaches, and beans) for more or less the length of time of the cold exposure. After prolonged cold, rewarming usually had a phase setting effect.

Because the phase shifts were not always the length of the chilling, the scientists did not think that the clock was "stopped" by the cold.[84]

Warm-blooded animals (the birds and mammals, homeotherms) maintain body temperatures around 37°C. There are circadian fluctuations in body temperature (about 1°C in a human, as much as 5°C in a house sparrow). Birds have higher body temperatures than mammals. Some homeotherms are able to drop their body temperatures on a nightly (*daily torpor*) or a seasonally (*hibernation*).[85]

Ablation

There are a number of surgical procedures that have produced arrhythmia, a loss of circadian rhythms, in constant conditions. Either the circadian pacemaker was removed. Or, alternatively, the measurable rhythm (the hands) were disconnected from the pacemaking oscillation (clock). Ablations which have abolished rhythms include removing the pineal gland from a house sparrow, lesioning the suprachiasmatic nuclei of the hypothalamus of the brain of a rat or hamster, removing the optic lobe from a cockroach. Thus, these structures—the pineal gland, the suprachiasmatic nuclei, and the optic lobes—are candidates for biological oscillators (circadian pacemakers, biological clocks).

Chemicals

Arrhythmia, or loss of circadian rhythms, has also been produced in organisms with various chemical or pharmacological treatments.

For example, the circadian locomotor rhythm of house sparrows disappears when they drink water that has melatonin in it, or when they are implanted with a capsule that constantly releases melatonin. Melatonin is a hormone made at night by the pineal gland and the retinae.

The circadian period length has been remarkably resistent not only to temperature, but to chemical influences. A *biochemical compensation*, similar to temperature compensation, has been proposed.

Sodium cyanide, arsenate, 2,4-dinitrophenol, and sodium fluoride reduced the amplitude of circadian rhythms.

Table 9.3 Chemical Alterations of Circadian Rhythms.[86]

Chemical	Alteration	Rhythm
Arsenate	Suppression	Sporulation, *Oedogonium*
CO_2 increase	Inhibition	Bean leaf movement
Colchicine	Transient long tau	Beans
Cycloheximide	Increase tau	Phototaxis, *Euglena*
D_2O, heavy water	Lengthens tau	Phototaxis, *Euglena*
D_2O, heavy water	Lengthens tau, 1 hour	Mammals
Dichlorodimethyl urea	Suppression	Photosynthesis, *Gonyaulax*
2,4-dinitrophenol	Suppression	Sporulation, *Oedogonium*
Estradiol	Shortens tau, − 7 minutes	Hamsters, *Mesocricetus*
Ethyl alcohol	Lengthens tau	Beans, *Phaseolus*
Ethyl alcohol	Lengthens tau	Isopod, *Excirolana*
Lithium	Lengthens tau	*Kalanchoe*
Melatonin	Arrhythmia	Activity, *Passer*
NaCN	Suppression	Sporulation, *Oedogonium*
NaF	Suppression	Sporulation, *Oedogonium*
Oxygen removal	Suppression	Bean leaf movement
O_2 replaced with N_2	Transient delay	Oat coleoptile growth
Puromycin	Inhibition	Luminescence, *Gonyaulax*
Testosterone	Lengthens tau	Activity, mouse
Theobromine	Lengthens tau	Beans
Theophylline	Lengthens tau	Beans
Urethane	Transient long tau	Beans

Chapter #10: Seasons and Photoperiodism

Calendar

Most of this book, so far, has dealt with daily or circadian rhythms, a biological clock. Many organisms also have a biological *calendar* that they can use to time seasonal events. Both operate at once. Think of a carousel. The horses are going up and down (daily cycle) and at the same time, the carousel of life is turning round and round (seasonal cycle).

We are not thought of as seasonal animals, in the sense that many other animals are seasonal breeders and their lives are tightly governed by photoperiod. Nevertheless, our schedules are certainly affected by time of year, and it is likely that our biology has seasonal changes because we evolved in, and adapted to, a world in which there were seasonal changes. With our clever use of fire and clothing, we were able to exploit latitudes with cold winter temperatures.

Daylength

We enjoy the spring progression of flowers. First the snowdrops march by, then the crocuses, next the daffodils, and the tulips bring up the rear. Most of us are aware that other events in nature correlate with changes in season, especially at the temperate and arctic latitudes. Temperature can provide seasonal information, but that information is less precise than the information provided by the daily changes in the times of dawn and dusk.

Daylength is affected by latitude. For example, at 60 degrees north latitude, the longest day of the year is about 18 hours and the shortest day of the year is about 6 hours.[87] At 30 degrees, the shortest daylength is 10 hours and the longest daylength is 14 hours. Near the equator, the light cycle is close to LD 12:12 throughout the year. In the arctic regions extremely long days and nights occur. At tropical latitudes in the vicinity of the equator, seasonal changes are less evident, and possession of a calendar may be less important to survival than in temperate and polar regions.

There is scant evidence that humans possess a biological calendar, but even prehistoric human cultures went to great lengths to make calendar calculations and may have erected monuments, such as Stonehenge, for astronomical predictions. It seems logical that calendar information had predictive value for agriculture and animal husbandry.

As I discussed, dawn and dusk provide cues as to time of day that can be used to reset circadian rhythms each day. At most latitudes, day and night length change precisely and systematically throughout the year. Thus the changing times of dawn and dusk—which methodically and predictably alter daylength (photoperiod) and night length (scotoperiod)—contain seasonal information as well as time of day information.

Table 10.1   Annual Changes in the Environment.[88]

Month	Temperature minimum, maximum, degrees Fahrenheit	Rain inches	Snow inches	Photoperiod
January	26,40	3.26	6.3	9 h 35 min
February	26,40	3.08	6.7	10 h 49 min
March	33,50	3.54	3.9	12 h 8 min
April	43,62	3.31	0.2	13 h 35 min
May	53,73	3.35	0.0	14 h 46 min
June	63,81	3.64	0.0	15 h 17 min
July	68,85	4.11	0.0	14 h 51 min
August	66,83	4.51	0.0	13 h 42 min
September	60,77	3.39	0.0	12 h 17 min
October	49,66	2.82	0.1	10 h 52 min
November	39,54	3.10	0.7	9 h 38 min
December	29,43	3.19	3.9	9 h 5 min

Photoperiodism

A response of an organism to the changes in day and night length is a *photoperiodic* response. The rationale that is offered is that there are adaptive advantages in being able to anticipate and make best use of the advantages, particularly of spring and summer for raising young, and to avoid the harsh winter conditions.

Experimentally, in the laboratory, it is possible for any given photoperiodic event to determine what duration of light (or dark) a short day (long night) is distinguished from a long day (short night). The phases *critical day length* or *critical photoperiod* refer to this.

Most measured critical photoperiods fall between 10 hours and 14 hours of light. The critical photoperiod within a species varies systematically with latitude. For example, the critical photoperiod for diapause in a butterfly, *Acronycta rumicis*, is 15 hours at 43 degrees north latitude, but longer, 18 hours, at 50 degrees north latitude.

For photoperiodism, light has a dual role: (1) it defines the photoperiod (and scotoperiod), and (2) it sets the phase of circadian rhythms. Some organisms seem to have separate physiological sites for the two functions. For example, daylength (or nightlength) detection resides in leaf blades, but the leaf joints detect light for phase shifting the circadian rhythm.

While it is popular to think of the "daylength" as the important physiological measurement being made, some investigators prefer to think that it is the length of the dark period (scotoperiod) that is actually measured, because disrupting the night with a light break spoils the measurement of day length. Some investigators consider the terminology in need of change so that a short-day plant should really be called a long-night plant, and a long-day plant is really a short-night plant.[89]

Table 10.2   Photoperiodic Events.

Appearance of sexual plant lice
Bulb dormancy induction in plants
Bulb dormancy termination in plants
Cambium activity
Diapause in insects
Flowering
Fur color change
Migration of birds

Table 10.2   *(Continued).*

Migratory restlessness
Molting
Reproductive cycles
Seed germination
Succulence
Testicular size
Tissue differentiation
Tuber formation
Vegetative development

Migration

A photoperiodic behavior that has intrigued biologists is animal migration. Animal migration depends upon season. Circadian investigators have proved that the circadian biological clock is used in direction finding, so these organisms have contributed to our perception of the biological clock as a *continuously consulted chronometer.*

The ability of pigeons to choose directions was studied by examining the individual vanishing bearings taken by the birds when they were released 30–60 km from their home loft. North, south, east, and west release locations were used. Normally all the birds would fly in the direction of their home loft. When the pigeons light–dark cycles were advance by 6 hours, shifting their circadian rhythms forward by 6 hours, the birds flew away in a direction that was 90 degrees in error from the direction of the home loft. This was taken as proof that the pigeons used their circadian clock in homing.[90]

The subject of bird migration has included intriguing questions not only about the biological clock, but also about magnetism and the use of the sun for celestial orientation. I have always wondered about a unique structure in the eyes of birds, named the pecten. The pecten is a heavily pigmented, usually black, comb-like structure that sits up at 90 degrees to the surface of the retina in some species of birds and reptiles.[91] The pecten looks for all the world exactly like the gnomen of a sundial. It should cast its shadow on the birds' retinas. I have wondered since I first saw it whether the birds can possibly use the pecten as an orienting device.

Reproduction

Season controls the physiology of many species. White-tailed deer mate in the fall, fawns are born in spring, the fur molts to reddish brown in spring and to gray in fall, and antlers are shed in fall and winter. Testis size changes amazingly with season in some species. The testes of juncos are very small in the non-breeding season (e.g. January) when the days are short. The juncos' testes grow dramatically when they are placed experimentally in long days.

Reproductive photoperiodism can be produced in the laboratory using Golden hamsters. Hamsters kept in long days (short nights) had large testes weighing at 3 grams a pair. But hamsters maintained in LD 8:16 had testes weighing only 0.4 grams a pair. The critical photoperiod for hamsters is precise, about 12.5 hours.

The male hamster's reproductive year goes like this: About the time of the autumnal equinox, when the days get shorter than 12.5 hours, the hamsters are sensitive to light-length or dark-length and the testes begin to regress. Since the hamsters are responsive to lengthening nights, they are said to be *photosensitive*. The testes not only get smaller, they stop producing sperm and hormones (testosterone). The testes remain small until March, near the spring equinox, and when daylength exceeds 12.5 hours they begin to enlarge (*testicular recrudescence*). At this time, if the hamsters are placed in constant dark, May for instance, their testes stay large; the hamsters are said to be *refractory*, or insensitive. About June, refractoriness terminates, and hamsters placed in the dark in June get smaller testes. It takes about 10 weeks for hamsters to enlarge or reduce their testes' sizes. Female hamsters also undergo starting and stopping of their reproductive cycles and are also controlled by the photoperiod so that they have much the same seasonal breeding periods as the males.

Light Break Experiments

The organism has a circadian rhythm of sensitivity to light. Day (or night) length is determined by when light (or dark) falls with respect to the underlying circadian rhythm of sensitivity. The duration of dark is the measured parameter. Light specifies photoperiod and also sets the circadian clock (*dual role*).

Evidence for the importance of dark-length comes from experiments with short pulses of light. In a typical light-break experiment, groups of organisms are exposed to asymmetrical 24 hours skeleton photoperiods in which there is a long light pulse (e.g., 6 hours) and a short light pulse (e.g., 2 hours). The short light pulses in the different treatment groups are spaced at intervals scanning the 18-hour dark period.

All of the skeletons contain the same total light-time, 8 hours. The 6-hour pulse synchronizes the circadian rhythm. The difference is when the 2-hour pulse falls. When the 2-hour pulses fall just after the 6-hour pulse (leaving a long night), the organisms show the short day response. When the pulses occur in the middle of the dark (e.g. 14–16 hours after D/L of the 6-hour light pulse), they show a long day response.

These experiments produce the same results with diapause in insects, with flowering in plants, and with testes of hamsters.

More evidence comes from still another kind of experiment using pulses called *resonance experiments* where the pulses are imposed at longer times after the entraining pulse. For example, in LD 6:18 hamster testes are small (long night, short day), they are large in LD 6:30 (perceived by the hamsters as LD 16:8 because light falls in their expected night), small in LD 6:42, and large in LD 6:54.

In both the light break and resonance experiments, the total light is less important to eliciting the testis response than *when* the light is imposed.

Photoperiodism, Melatonin, Pineal Gland, and Eyes

An endocrine explanation has been possible for the photoperiodic control of reproduction. The explanation involves the pineal gland. The pineal gland is in the center of the head of humans (surely important!) but rises from the center to just beneath the skull in other species, such as sparrows. The pineal (pine-eel) gland is named for its pine cone shape. The pine-cone symbolizes fertility, regeneration, healing, and conviviality. Osiris and Attis were spirits of pine trees and Dionysus carried a thyrsus with a pine cone at its top. While

Whitten writes that the pine-cone had "prophylactic qualities," I think he means healing in general, not a recognition of the fact that the pineal makes a hormone, melatonin, that inhibits the reproductive system.[92] The pineal gland has suffered the misfortune of being referred to as the penis of the brain.[93] Some folks thus say peen-yeel.

Removing the pineal gland[94] prevented the dark-induced regression in hamster testes.[95] The pineal gland makes a hormone, *melatonin*, that inhibits the reproductive systems of hamsters. Melatonin is also made by the retinas of the eyes.

Melatonin production in the pineal gland is regulated by photoperiod.

Melatonin production is controlled by two enzymes. An indole molecule in the pineal gland, named serotonin (5-HT) is the precursor molecule. Using acetyl coenzyme A, the enzyme *N-acetyltransferase* (NAT) acetylates the serotonin producing N-acetylserotonin. The enzyme *hyroxyindole-O-methyltransferase* (HIOMT) uses S-adenosyl-methionine to methylate N-acetylsertonin producing melatonin.

serotonin $\longrightarrow$ N-acetylserotonin $\longrightarrow$ melatonin

The melatonin is secreted, or released as a hormone, and carried by the blood. Melatonin most likely achieves its action on the reproductive system indirectly by acting on the hypothalamus of the brain to influence the production of prolactin, a gonad-stimulating hormone.

melatonin $\longrightarrow$ prolactin inhibition $\longrightarrow$ gonad regression

The N-acetyltransferase has 30-fold rhythms of activity that respond to light and dark, and thus it controls the daily cycle of melatonin production, both its night time peak, and the duration of that peak. Constant light (LL) suppresses the amplitude of the NAT cycle; the NAT cycle freeruns in constant dark (DD). HIOMT also responds to light and dark in some species and may play a modulating role in seasonal control.

Thus, the sequence becomes:

long night $\longrightarrow$ N-acetyltransferase $\longrightarrow$ melatonin $\longrightarrow$

prolactin inhibition $\longrightarrow$ gonad regression

Dark cannot stimulate NAT at just any time, but only curing a sensitive period that occurs in the night. Light causes a rapid decrease (plummet) in NAT activity once it is high. This sensitivity to light pulses at night, reducing the apparent length of the dark time, and blocking long melatonin synthesis, explains light break and resonance experiments. The inactivation of NAT is so rapid, that it might be best explained by formation or dissolution of a dimer.

Genes provide the sequence of amino acids that are used to make proteins of which NAT, an enzyme, is a protein. Some inbred laboratory mice have a deficiency in the ability to synthesize melatonin which is believed to result from mutations in two genes which may be responsible for absence of normal NAT and HIOMT activity.[96]

Chapter #11: Circarhythms

There are four rhythms which correspond to and synchronize with geophysical cycles in the natural environment. Collectively, they are referred to as the circarhythms. These four rhythms persist, they entrain, and they are temperature compensated. These are not the only biological rhythms that have been named and studied. The four circarhythms correspond to four environmental rhythms. There are other biological rhythms for which there is no corresponding geophysical cycle.

tides	circatidal
day and night	circadian
phases of the moon	circalunar
season	circannual

For us, the circadian cycles are most relevant, with circannual cycles in second position.

Short Cycles

An *ultra*dian rhythm, or high frequency rhythm, has a period shorter than 24 hours, or a frequency greater than one cycle in 20 hours. This would seem to include circatidal rhythms, and bimodal patterns of activity in house sparrows, but they are usually treated as a separate case. The "ultra" refers to frequency, not period length.

There are 24 eyeblinks/minute in the human. Compound action potentials (CAPs) have up to 300 impulses/hour in the optic nerve of the *Aplysia* eye. 12–15 respirations take place every minute and there are about 70 heartbeats/minute in the human. Erections take place every 90 minutes in sleeping men.

The pacemaker for the heartbeat resides in is in the heart's sino-atrial node. The pacemakers signals are transmitted electrically to other parts of the heart. Pieces of hearts are capable of beating when isolated *in vitro*, but the fastest rate is in the sino-atrial node. The signal from the fast beating node drives the rest of the heart. The heart has a pacemaker, a driving oscillator, and its beat might be considered to freerun. However, there is not an environmental signal that entrains the heartbeat. In the laboratory, or with an electronic pacemaker, the heartbeat can be driven experimentally, and shows the phenomenon of frequency demultiplication. Hearts do not exhibit temperature compensation, the turtle heart rate speeds up as temperature increases, and it slows down when the temperature decreases. Signals transmitted to the heart via the vagus nerve slow it down. A transplanted heart is dependent upon its own pacemaker.

Siberian hamsters (*Phodopus sungorus*) exhibited a circadian rhythm of wheel running activity in constant light or dark. However, the hamsters did not run steadily. They ran the wheels in short bouts, usually less than one hour per bout. Most days, a typical hamster had six bouts of activity, but occasionally it had five or seven. There is no known Zeitgeber for the bouts.[97]

It was possible to drive sparrow activity with short period light dark cycles, such as LD 1.5:1.5, a 3-hour cycle. Sparrows exposed to 10-minute cycles with 3 minutes of light were active in the light producing a "staccato" pattern. When the 10-minute cycle contained only half a minute of light, the sparrows were not always active in the light. This may mean that the direct effect of light needs more than half a minute, and three minutes is enough to evoke it. Sparrows in the 10-minute cycles exhibited circadian patterns as well as those whose period lengths were shorter than their freerunning periods in constant dark (DD). The 10-minute and 3-hour cycles were imposed artificially in the laboratory. There is nothing like them in natural conditions.

Sleep and 90 Minutes

Sleep researchers have provided us with an understanding of sleep as an active process. They have variously divided human sleep into stages. For example, in one scheme, individuals are considered to be awake, in REM sleep (rapid eye movement sleep), or in NREM (non-REM sleep including sleep stages 1–4). Classification of sleep states is based on recordings from the eye (EOG, electrooculogram), chin (EMG, electromyogram), and brain (EEG, electroencephalogram). REM sleep is called paradoxical sleep because some of its EEG characteristics are more similar to wakefulness than to other states of sleep. At the same time, it has been more difficult to awaken REM people from REM sleep than when they were in other sleep stages. So despite the EEG characteristics, REM sleep is deep sleep. Typically, REM sleep has been characterized by rapid eye movements, dreaming, penile erections, a drop of tonic chin muscle activity, depression of spinal reflexes, and low amplitude EEG waves.

The occurrence of REM sleep has been mainly documented in mammals. Something akin to REM sleep has been observed in birds, and its occurrence in lizards and other lower vertebrates is questionable. The reason we are discussing REM sleep here is that it has an ultradian rhythm; in the human it recurs through the night in cycles that are 70–90 minutes long.

In mammals other than humans, the frequency of REM sleep increases as body weight decreases. Mice, for example, have a 7-minute cycle. Humans may have a basic rest-activity cycle (BRAC) for all 24 hours of which REM sleep is a night-time manifestation. The BRAC cycle is 10–13 minutes in the rat, 30 minutes in the cat, 45 minutes in the monkey, and about 120 minutes in the elephant.[98]

REM sleep may depend on mechanisms involving serotonin, the caudal part of the raphe system of the brain, cholinergic and noradrenergic mechanisms, and the locus ceruleus of the brain.[99] Drugs affect REM sleep. For example, one drug, para-chlorophenyl-alanine (PCPA) inhibits synthesis of serotonin in the brain and induces insomnia.

Estrous Cycles

The time domain of rhythms with longer than 24-hour periods, or lower frequencies than circadian rhythms, is occupied by *infradian rhythms*.

Probably the most studied infradian rhythms are the reproductive cycles that are found in female laboratory rodents—mice, rats, and hamsters. These animals enter and exit a period of sexual receptiveness called heat, or estrous, during which they are willing to mate and able to conceive. Estrous is correlated with and caused by a rhythm of estrogen from the ovaries. Estrous in the rat occurs during the dark-time for a few hours every fourth day. Correlating with the

4-day heat cycle are other events such as changes in hormone levels and vaginal cytology. Estrous cycles persist in constant conditions. But there is not a 4-day Zeitgeber. Constant bright light abolishes the cyclicity in rodents and they go into permanent estrous. Removing the ovaries leaves the remaining reproductive system in a quiescent state, diestrous. The hypothalamus of the brain is the likely location of a 4-day clock sending endocrine signals to the ovaries by releasing gonadotropins (follicle stimulating hormone or FSH, luteinizing hormone or LH, and prolactin) from the pituitary gland.

Estrous cycles vary in length, depending on the organism.

Table 11.1   Estrous Cycles in Some Mammals.[100]

Organism	Cycle Length (days)
Cow	21
Goat	20–21
Sheep	16
Pig	21
Horse	19–23
Dog	60
Mink	8–9
Fox	90
Ground squirrel	16
Guinea pig	16
Golden hamster	4
Mouse	4
Rat	4–5

Lunar and Tidal Rhythms

"And God made two great lights; the greater light to rule the day, and the lesser light to rule the night."[101] Despite this verse from Genesis, the moon is actually there all the time, day and night.

All the organisms on earth are subjected to cycles from the moon's rotation about the earth. The organisms that live in the seas have the most easily observed effects of the moon because of the tidal cycles that are caused by the gravitational pull of the moon. The lunar day is 24.8 hours (24 hours and 51 minutes).

There are two tides per lunar day, recurring at 12.4-hour intervals. The time between two high tides is called the *tide day*.

I was raised in the midwest, and didn't have to pay any attention to the tides. Only later in my life, when I had a chance to spend time in Stone Harbor, New Jersey, and to see the enormous tides in Digby, Nova Scotia, did I have a chance to appreciate the way the moon sloshes the oceans back and forth.

Why is the period longer than 24 hours? The moon circles the earth in the same direction that the earth rotates on its axis. The earth takes 24 hours to rotate relative to the sun. The moon's travels have meanwhile placed it in a new location. To face the moon again, the earth must play catch-up and rotate an additional 13 degrees.

The amplitude and timing of the rhythm of the tides vary with longitude (large tides towards the poles), phase of the moon, and wind. The rising tide is the *flood* tide. The falling tide is the *ebb* tide. The unusually high tides at the full and new moons are the *spring* tides. The unusually low tides at the first and third quarter moons are the *neap* tides. And a tide running against the wind is a *weather* tide.

A beach organism, then has a number of time cues available that recur in each 24.8-hour period—dawn, dusk, two high tides, and two low tides.

Dual rhythms, both daily and lunar, have been simultaneously measured in individuals under natural conditions. The penultimate hour crab (*Sesarma*) exhibits both a 24-hour and a 24.8-hour component in its activity rhythm.[102] And a fiddler crab (*Uca*) kept in the laboratory in a natural light–dark cycle displayed a tidal rhythm with beats at 24 hours.[103]

The commuter diatom, *Hantzschia*, has a vertical migration rhythm. The organism is a microscopic alga that inhabits the sand. At high tide, the organism descends among the sand grains, but at low tide, the organism rises to the surface in sufficient numbers to give the beach a golden brown color.

Various factors associated with the tides may act as cues to synchronize the organisms' tidal rhythms. Inundation (wetting) does not seem to be important, but temperature, pressure, and mechanical agitation can act as cues in the laboratory.

It is not only aquatic animals that have lunar and tidal rhythms. The ant lions build larger pits when the moon is full, and smaller pits when the moon is new.[104]

Lunar and Menstrual Cycles

Imposition of a lunar day is not the only periodic signal provided by the moon. There is also a light–dark cycle provided by the waxing and waning of the moon that produces a lunar month (a synodic month, a lunation) of about 29.5 days. Monthly variations have been found in flatworm (planarian) phototaxis and bean water uptake that may synchronize to the lunar month. The lunar month has been subdivided into eight phases: new, crescent (waxing), first quarter (half-moon), gibbous, full moon, gibbous, last quarter (half-moon), and crescent (waning).

The human menstrual cycle averages 29.5 days between the ages of 15 and 40 years. Cycles tend to be longer than 28 days (the mode length of the menstrual cycle) at puberty and menopause.

Is the menstrual cycle a circalunar rhythm? The variation in period length among and within individual women argues against a lunar cue in modern society.[105] However, we mask the lunar cycle with by adding home lighting and by blinding our windows. Dewan attempted to synchronize women's menstrual cycles by keeping on the lights at night for days 14–17 after the first day of the last menstrual cycle, and claimed that the treatment regularized the women's cycles toward 29.5 days.[106]

Never mind the moon, is there social synchronization for the menstrual cycle? McClintock argued that reduced individual deviations from the group mean time of menstruation in a college dormitory over the course of a 9-month academic year was evidence for social synchronization of menstrual cycles among women housed together.[107]

Even if modern women's menstrual cycles are not synchronized by the moon, one might speculate that the period is derived from an ancestral timing that evolved when our foremothers lived under more natural conditions.

Still Longer Cycles

Perhaps the most amazing biological cycles are those that are even longer than a month in length. From January to May, but not the rest of the year, oxytocin is released in goats.[108] Serotonin fluctuates with two cycles per year in *Ammocoetes*, the larvae of lampreys.[109] Such twice-a-year cycles have not been studied much, but a great deal of attention has been paid to annual cycles, the circannual rhythms. Seasonal cycles were already discussed in this book, particularly in connection with seasonal control of reproduction and photoperiodism.

Here, the emphasis is on the fact that some seasonal cycles have been shown to persist in constant conditions. These studies require dogged persistence by an investigator.

Cycles of fattening and hibernation[110] have been shown to persist for over three years (three cycles) in ground squirrels. Body weight cycles over a hundred grams, almost half the squirrels' weights, between the onset (when the squirrels are fat) and the end (when the squirrels are thin) of hibernation in individual golden-mantled ground squirrels (*Citellus lateralis*). Deprived of light and at 3°C the squirrels' hibernation cycles freeran with periods ranging from 293 to 369 days in length. Such cycles are evidence of circannual rhythms or clocks.

Most of the cycles measured were shorter than the 365 and a quarter day that the earth takes to complete its circumsolar path. Other observed circannual rhythms include: growth and longevity in coelenterates, oviposition in mollusks, behavioral thermoregulation in fish, and activity in reptiles. Molt, migratory restlessness, body weight, fat deposition, and gonad size have circannual rhythms in birds. Molt, plasma androgens, reproductive condition, nest building, locomotor activity, antler replacement, milk production, and water consumption have circannual rhythms in mammals. Usually, the organisms were held in unchanging light–dark cycles to prove the rhythms were circannual. For example, molting cycles were measured in chickadees and warblers kept in LD 10:14 for 10 years.

Some factors that influence the duration of circannual rhythms have been discovered. The duration of the annual cycle of body weight in golden-mantled ground squirrels because their average circannual period lengths were two months longer in 9.5°C than in

21°C. Social isolation reduced the period length of the testicular size cycle in starlings. Lesions of the hypothalamus or pinealectomy shortened the period length in ground squirrels.[111]

Is there a Zeitgeber(s) for the circannual rhythm? Since circannual freeruns are generally not exactly a year, some event(s) must bring the seasonal cycle into synchrony with seasonal events in the environment. The answer so far seems to be that a multitude of factors may be involved in synchronizing circannual rhythms.

There is evidence for photoperiodic regulation of seasonal reproduction, so clearly photoperiod is involved. Resynchronization was shown by transferring woodchucks from Pennsylvania to Australia, which reversed their annual cues, and reversed (reset) their body weight cycles. An annual cycle of photoperiod can synchronize a circannual rhythm of molting in starlings. T cycle experiments can shorten starling circannual rhythms to as little as 2.4 months. The range of entrainment for the sika deer antler replacement cycle is 4–24 months. Transfer to warm or cold temperature can phase shift the circannual rhythm of hibernation in ground squirrels. Rams may be able to transmit seasonal information to ewes.

Do humans have circannual rhythms? or even seasonal rhythms? Before exposure to modern civilization and lighting, Eskimo women supposedly stopped menstrual cycles in the winter. Eskimo passions were claimed to be periodic, with courtship soon after the return of the sun in the arctic spring.[112]

Seasonal changes in melatonin, conceptions, testosterone, and menstrual cycle length have been reported. In Finland, where daylength varies greatly because of the high latitude, summer (June to September) had the highest conception rate, and the conception rate was lowest in winter (November to February).[113]

Chapter #12: Biological Clockworks

We can tell the time on a grandfather clock from the position of the hands on the clock face. We can open the back and front of the grandfather clock and look directly at the mechanism of wheels and weights that determine the position of the hands. That is, we can look at the clockworks.

Discovery of the clockworks of living organisms has required more detective work.

The human pacemaker is likely in the suprachiasmatic nuclei, and the human circadian system is a hierarchy of oscillators including the pineal and pituitary and adrenal glands.

But, where, and what, are the clockworks that produce circadian rhythms?

Black Boxes

Scientists played "What if?" trying to guess how the biological clock might work. They treated the biological clock mechanism as a *black box* and attempted to create clockworks using what they knew about the properties of measured rhythms, rules of science, and biological components.

Hourglass models are simple. We all understand how the sand or water—water clocks, clepsydra—flows through the tiny hole of the infinity shaped glass bottle at a constant rate and measures an interval. Indeed, hourglasses are literary and artistic symbols for time passing. To make a rhythm, though, the hourglass has to be tipped. Still, the hourglass principle may be used to measure some rhythms such as the timing of photoperiodic responses in aphids.

And we all know how a *pendulum* works. So we can think of a child on a swing as the freerun and the pushes needed to keep the child swinging as the Zeitgeber and entrainment.

If you worked in a laboratory, you might have seen a pipet washer which is an example of a *relaxation oscillator*. Water from a faucet runs through a tube continually into the top of the washer. As the washer fills, so does a u-shaped pipe on its side. When the water level reaches the top of the pipe, it overflows so that a siphon is created. The siphon, which is a larger diameter than the input tube, removes the water from the washer at a faster rate than its inflow from the faucet. Once the washer is emptied, the siphon is interrupted and the washer is again slowly filled by the water from the faucet. The period of oscillation can be controlled by adjusting the faucet handle.

Hourglass, pendulum, and relaxation oscillators can be described with mathematical formulae. When this was done, the wave forms, solutions of the van der Pol equation, ranged from sine to square waves. The wave forms of measured biological rhythms, such as sparrow freeruns, were very nearly sinusoidal.

Builders also visualized the clock as a circle with a dot in the center, representing a *limit cycle* or *attracting limit cycle*. In their models the circadian clock is viewed as a loop (a graph of pendulum velocity). A cycle of a freerun represents a course, or trajectory, around this loop. Time positions on the loop were called *isochrons* and the center was called the *singularity*.[114] The clock was visualized as being started by an LL/DD transition. Perturbing stimuli, such as light pulses, move the position along its trajectory, producing resetting. If the clock is forced to the center of the circle, the result is arrhythmia or a jump to an unpredictable new phase, or position on the trajectory.

Single cells are probably capable of generating circadian rhythms (swimming activity, phototactic sensitivity, bioluminescence, photosynthetic capacity, sexual reactivity) simply because single celled organisms (*Euglena, Gonyaulax, Paramecium, Chlamydomonas*) have circadian rhythms. Measuring rhythms of these organisms has been technically feasible using populations, and not individuals, so there is some possibility that the rhythms are produced by mutual interactions. Single *Gonyaulax polyedra*[115] had oxygen rhythms in LD and LL and single *Acetabularia* had rhythms. Populations of chick pineal cells in culture have rhythms.[116]

In multicellular organisms, the circadian rhythms of individual cells are visualized as being *coupled* so that they are synchronized.

The many various rhythms of inside individual organisms, like ourselves, are believed to be organized into *hierarchies of oscillators* consisting of *pacemakers* (driving oscillations) and *slaves* (driven oscillations). A sequence of events is imagined: The time cues

(Zeitgebers, light) are detected (by photoreceptors). The environmental information is then relayed (by nerves) to a pacemaker (a group of nerve cells) which in turn drives the measurable driven rhythms (hormones, activity, body temperature).

light $\longrightarrow$ photoreceptor $\longrightarrow$ pacemaker $\longrightarrow$ driven oscillations

A *two oscillator model* was proposed in which there is one oscillator that is *locked* to, or synchronized by dawn, M, a morning oscillator. The second oscillator is locked to dusk, the E, or evening oscillator. This model seems to account for such phenomena as "splitting" of hamster rhythms in LL and "bimodal" activity patterns in house sparrows kept in LD.

It is possible to account for changes in the shape of an oscillation with a *one oscillator model*. The circadian oscillation (imagine a sine wave) has a mean (average) value that interacts with a *threshold* (imagine a horizontal line running through the sine wave). The observed rhythm is determined by how much of the oscillation is above the threshold. By raising and lowering either the threshold or the mean, the amount above threshold is altered and with it the duration and amplitude of the portion above the threshold. The model explains the effects of light intensity on the alpha/rho ratio and the changes in the shape of an oscillation observed in different photoperiods.

Pacemakers

The method that has been successfully used to locate pacemakers and photoreceptors has been ablation. Removing a suspected area can be done many ways, with surgery or lesions, with chemicals, and sometimes with mechanical means, such as a hood to block light from reaching a photoreceptor.

The method was successfully used to work out the sequence of structures involved in the circadian system of the rat:

dark $\longrightarrow$ eyes $\longrightarrow$ retinohypothalamic tract $\longrightarrow$

suprachiasmatic nuclei $\longrightarrow$ superior cervical ganglia $\longrightarrow$

pineal N-acetyltransferase increases $\longrightarrow$

In this pathway, the eyes are the photoreceptor and the supra-chiasmatic nuclei (SCN) are the location of the pacemaker. The connections are all nerves. In the pineal gland, the nerve signal, received as norepinephrine released by endings of nerves whose cell bodies are located in the superior cervical ganglia (SCG) in the neck. The norepinephrine (NE) stimulates the activity of the enzyme N-acetyltransferase (NAT) causing an increase in melatonin production. This conversion of a neural signal (NE) to a hormone increase (melatonin) is *neuroendocrine transduction*. The melatonin is picked up by the blood and carried all over the brain and body where it can influence other physiological events.

Pacemaking function was suspected of structures whose removal causes a loss of circadian rhythmicity. Putative pacemakers were qualified by isolating them and determining whether or not they can create a circadian rhythm on their own. The pineal glands of birds have pacemaking properties, as do the suprachiasmatic nuclei of vertebrates. In the case of the pineal gland, its hormone melatonin has been shown to synchronize rhythms when injected every day, and constant melatonin administration has caused arrhythmia.

Cell Cycles

Within cells of tissues, such as rat liver, there are visible daily changes in the volume and number of cell nuclei. Binucleated rat liver cells comprise 9% of the population in the light and 18% in the dark.[117] The peak times of cell division (acrophase of mitoses) reported have varied, but, for example, may occur at one time of day, such as in the early morning at the time of lights-on in nocturnal mice.[118]

If single cells have their own rhythms, it follows that all of the clockworks must be contained in some cells. A cell cycle clock (cytochron) was proposed when it was noticed that there were correlations between the cycle of cell division and other circadian rhythms. Cell division rhythms persist and cell division occurs in the subjective night. The cell cycle is represented with abbreviations (given below with their meaning)

$$G_1 \longrightarrow S \longrightarrow G_2 \longrightarrow M$$

or, written differently,

gap 1 $\longrightarrow$ chromatin replication $\longrightarrow$ gap 2 $\longrightarrow$ mitosis

In the model, a cytochron timer starts at lights-on and the stages $G_1 \longrightarrow S \longrightarrow G_2$ proceed and trigger M and cytokinesis.

Chronons and Genes

A *chronon model* for the circadian clock makes use of the sequence of events by which proteins are synthesized in cells. A sequence of chemical bases in the genetic material in chromosomes of the nucleus (DNA) specified a sequence of complementary bases in the formation of a large molecule (RNA), a process called transcription. In turn, the RNA leaves the cell nucleus and provides instructions for forming the sequence of amino acids in proteins at the cellular organelles called ribosomes, a process called translation. A cell contains chronons, strands of DNA 200–2000 cistrons long, which are the rate-limiting components of transcription, and therefore provide the timing mechanism. In this model, transcription takes a certain duration of time at the end of which some substance synthesized at the ribosomes then diffuses back to the DNA to restart the cycle. The role of diffusion in this model provides an explanation for temperature compensation, because diffusion has a Q_{10} near 1.

Membranes

A *membrane model* is based on the structures of cell membranes, potassium ions, and the circadian Q_{10} evidence for a diffusion process. In this model cell membranes are viewed as consisting of a lipid bilayer studded with nuggets which are moveable proteins. Changes in this *fluid mosaic membrane* determine the direction of activity of transport so that the potassium ion concentration, whether it is high inside or outside the cell, oscillates.

N-Acetyltransferase

Model-making is fun, and anybody can do it, and I did too. I built my model for an "enzyme clock" using N-acetyltransferase (NAT), the enzyme of the pineal gland and the retina, that I spent so many years studying in chickens. The activity of NAT increases at night, resulting in increased melatonin. NAT is inhibited by light. Unexpected light at night results in a rapid plummet in NAT activity, NAT is low in the day, and constant light abolishes the rhythm of NAT.[119]

frequency in **Bread Mold**

A model, based on molecular biology, has been offered to explain the way that light achieves entrainment of circadian rhythms of *Neurospora*. The idea is that the light-induced resetting of the circadian clock, which I have already said is rapid and instantaneous, is caused by a rapid increase in the transcript of *frequency* (*frq gene*). First, light causes a 4–25 fold increase in 15–30 minutes in the mRNA produced from the gene *frq*. In turn the mRNA should produce more protein, FRQ. The threshold for the effect is 8 micromol photons/square meter and it takes 10 micromol photons/ square meter to reset the clock. Constant light keeps *frq* mRNA high, but as soon as they are placed in dark, *frq* transcript drops to low levels.[120] The FRQ may down-regulate the *frq* mRNA, a negative feedback, which creates the circadian oscillator.[121]

light $\longrightarrow$ *frq* $\longrightarrow$ *frq* mRNA $\longrightarrow$ FRQ

light $\longrightarrow$ *gene* $\longrightarrow$ messenger RNA $\longrightarrow$ protein

period and *timeless* in **Flies**

A model, based on molecular biology, has also been offered for fruit flies using genes named *period* and *timeless* (*tim*).[122]

Timeless, tim gene $\longrightarrow$ *tim* mRNA $\longrightarrow$ TIM protein

Period, per gene $\longrightarrow$ *per* mRNA $\longrightarrow$ PER protein

PER accumulates in the cell sap (cytoplasm) until something (unknown) causes it to move to the cell nucleus. It can't make this migration without TIM, so the two proteins, PER and TIM may bind to one another. Once PER is in the nucleus, it inhibits the *per* gene expression so that PER production is reduced—a case of negative feedback.

A temporal sequence is imagined:

c.t. 0 (dawn) *per* begins to make PER.

c.t. 12 (dusk) *tim* begins to make TIM. The TIM and PER form dimers. The dimers migrate to the nucleus.

c.t. 18, the dimers are in the nucleus, and they stop the expression of the *per* gene so that PER production stops.

To make the system work as an oscillator, there is some means of getting rid of the dimers—breaking them down (turnover without replacement), dissociating the dimers, or consuming the dimers in the process of turning off *per*.

Bestiary

The black box was opened and real biological clocks were found.

In rats (*Rattus norvegicus*),[123] the pacemaker is in the hypothalamus. Lesions of the hypothalamus of the brain, particularly of the **suprachiasmatic nuclei** (SCN), caused the rats to lose their freerunning circadian rhythms, that is, they were arrhythmic. The same result was obtained in hamsters and other species, leading to the conclusion that the circadian pacemaker, the biological clock, in mammals, including humans, is located in the hypothalamus in two clusters of cells called the suprachiasmatic nuclei. The hierarchy described above applies in mammals and probably in humans with the additional information that the retina also synthesizes melatonin.

Pinealectomized sparrows lost their circadian rhythms in constant dark (DD). Chick, sparrow, and lizard pineal glands can produce melatonin rhythmically when the pineal glands are isolated in organ culture, which means they are capable of generating circadian oscillations. Pinealectomized sparrows receiving pineal transplants, get their freeruns back with characteristics of the donors' freeruns.

Removing chicken pineal glands hasn't abolished their rhythms, maybe because their eyes make as much melatonin as their pineal glands, but chicken pineal glands are good performers in organ culture, exhibiting timing in their N-acetyltransferase activity, and able to produce melatonin in daily cycles.

Cockroaches, as other invertebrates, have neither pineal glands nor suprachiasmatic nuclei. However, they do run wheels and their rhythms were studied. The pacemaker is located in the cockroaches' optic lobes. Transplanting optic lobes restores rhythmicity. Although the system seems very different, roaches have serotonin, NAT, HIOMT, and melatonin.

Sea hares (nudibranchs, *Aplysia californicus*), marine gastropods (*Bulla bulla*), and pulmonate slugs (*Limax pseudoflavus* Evans) have nervous systems that contain giant neurons, nerve cells a millimeter in diameter. Rhythms are measurable in whole mollusks using wheels or trippers suspended around the tanks, but it is also possible to record circadian rhythms of electrical activity in the eyes. The eyes of *Aplysia* are small, barely visible, 0.7 mm dark pips. Circadian rhythms have been recorded from isolated eyes and they are thought to contain both photoreceptor and pacemaker function. The eye rhythms freeran (tau = 23.9 to 27 hours), shortened in constant light, were phase shifted by light pulses, and were temperature compensated. Here again, serotonin is a "character," having been found to be present and also to shift the phase of the eye rhythm in a manner that correlates with dark pulse PRCs discovered in other animals.

Chapter #13: Human Rhythms

The application of circadian principles to animal husbandry and agriculture has been practiced for so long that we take it for granted. Use has been made, even by primitive cultures, of knowledge of the seasonal habits of plants and animals. It is possible to control the timing of fertility of economically important species by manipulation of the photoperiod. The breeding activity of ewes can be advanced by 10 weeks using either short photoperiod[124] or melatonin.[125] Chickens are included to increase their egg production and growth with constant light. Planning for seasonal changes in weather and light is necessary for success in raising most crops. If you are a gardener, you have probably made firsthand use of planting guides for your vegetables, and you probably have expectations of the sequence in which your flowers will bloom.

Here, we consider the application of circadian rhythm principles to human endeavors. We know how circadian systems work, and we can make intelligent suggestions for coordinating your activities with the timing of the physiology of your body.

I have studied human circadian rhythms, and the circadian rhythms of a variety of species in the laboratory, and there is an overwhelming observation that is not apparent from looking at any one record. The circadian behavior of a group of 20 sparrows placed in a natural light–dark cycle are much more similar to one another than the circadian behavior of a group of 20 human subjects living on their own and going about their normal lives. Most of the sparrows exposed to natural dawn awaken within a few minutes of one another. Someone once joked, "Birds of a feather clock together." Sparrows are synchronized!

Human subjects we studied, in remarkable contrast, awakened, on the average, between 5 a.m. and 10 a.m., having as much as five-hour differences from one another (and that's not the extremes, just the averages!). In humans, individual differences in circadian behavior are the norm. Thoreau noticed that we do not all keep pace with our companions, but we step to our own music.

Entrainment

We take for granted the organization of our time into a waking day and a sleeping night. This makes us a basically diurnal species, however, we are much different from house sparrows, which rise at dawn and roost at dusk, because of our extension of our activity into the dark-time with artificial light.

Somewhere along the way a rainmaker grabbed a fistful of lightning, and zinged a thunderbolt to the ground, and started a fire. Perhaps you have not thought about the dramatic effect this probably had on our biology in that it meant light as well as heat, but since we learned to use fire and became firestarters, we have not only obtained the use of heat, but the extra light.

Physicians and scientists have catalogued a host of physiological measurements that have daily cycles. Peak values of these daily cycles are not all at the same time. Peak values occur in urine excretion in the early morning, mental state and physical vigor in the mid-day, weight in the late day-time, deaths from surgery and air accidents in the middle of the night, and births in the late night. We are most sensitive to pain around 5 p.m. and we metabolize alcohol best between 2 p.m. and midnight. Alcohol consumption in the evening thus should be less intoxicating than at 6 a.m.

As pointed out previously, in practice we make use of alarm clocks which we set ourselves, and differ from other species in that respect. By using watches and clocks, we closely time events in our lives. Presumably, the ability to synchronize with other individuals by use of precise timing devices as been advantageous in our evolution.

Still, we can wonder whether our circadian rhythms can entrain using the same cues established for plants and animals—light, temperature, and social cues.

Human melatonin rhythms respond to light and to photoperiod. In humans, as in animals, melatonin is high at night and low in the day (measured in urine or blood). Light at night suppresses human melatonin, and the human melatonin wave form varies with season of the year.

Light is an effective Zeitgeber for humans under experimental conditions.[126,127] Light pulses shifted the phase of human circadian rhythms.[128,129] It was possible to entrain humans to artificial 27- and

29-hour light–dark cycles (T cycles)[130] and, as mentioned before, submariners sometimes use an 18-hour T cycle.

Social cues were claimed important for human entrainment in one study, where subjects remained synchronized during four days of complete darkness.[131]

It is of practical interest to parents to look at the development of the sleep–wake pattern in human babies. For the first month or so, babies don't have a circadian rhythm, and if anything there seems to be a freerun of about four hours. After the eleventh week or so, the circadian rhythm is established and synchronized with the day and night regimen.[132]

Some abnormalities have been noticed in the rhythms of blind human subjects, such as freeruns or an unusual phase of the melatonin rhythm.[133] The evidence from blind people appears to confirm the idea that there is some role for light in human circadian rhythm regulation. However, many blind subjects synchronize normally compared to individuals with sight.

Humans have rhythms with periods longer than 24 hours. One example is the menstrual cycle. Annual fluctuations occur in births, especially at the higher latitudes. Ultradian oscillations have been observed in infants and the occurrence of paradoxical sleep (See Circarhythms).

Weekend Delay

Although I think of people as "entrained" to their 24-hour environment (excepting people who travel across time zones, or do shift work, work in submarines, or volunteer to live in bunkers or caves), we have a peculiar pattern of entrainment that is certainly a consequence of our weekly work schedules. University personnel were found to have lower heart rate averages and different circadian peak times than did factory workers and both groups had lower heart rate averages than air traffic controllers.[134]

I analyzed the self-reported times of retiring and arising in humans using the same technique described in Section III of this book. I studied the records that were kept by 11 graduate students for a month in the autumn, and we found that the students exhibited a weekend delay in wake-up time of about 2 hours, and a

Monday-Wednesday-Friday effect. I explain these changes by the absence of weekend classes, and the fact that weekday classes meet MWF or TTh.

I kept a longer record myself, and included information as to which days alarms were set. When the alarm days were removed, the record was erratic, but the pattern still was characteristic of entrainment with a 24-hour cycle and it was not freerunning. In contrast, the onsets of wheel running activity of flying squirrels kept in natural lighting are closely associated with dusk, and sparrow perching activity in natural light starts near dawn and ends near dusk.

I analyzed data collected by six different groups of students over the years 1989–1995. The groups had 11 to 35 participants each (a total of 255 altogether), and each participant collected data for a month—about 2650 person-days of data overall! The average weekend delays—the latest minus the earliest wake-up of the week—varied from 1.9 to 2.6 hours for the six groups. The earliest wake-up day of the week in the six studies varied: Monday, Tuesday, Tuesday, Wednesday, Thursday, and Friday. The latest wake-up day was on the week-end, Sunday in five of the studies, and Saturday in the sixth. The earliest to-sleep day varied: Monday, Monday, Monday, Tuesday, Thursday, Sunday. The latest to-sleep days preceded the days when there were no classes: Saturday in four studies and Friday in the other two. The day on which people were awake the longest was Friday (17.2–18.2 hours) in five studies, and 16.6 hours on Tuesday in the sixth. The shortest awake day was invariably Sunday (14.0–15.7 hours).

So we have 24-hour rhythms, and we also have weekly rhythms.

Freeruns in Caves

Freerunning rhythms have been studied in humans by depriving them of time cues. This was done in caves, in bunkers, or in specially designed laboratories. Usually the subjects were isolated individually and kept at constant temperature. Sometimes the individuals were permitted to choose the lighting. Usually they had no watches or clocks. Rarely, however, were they kept in constant dark, DD.

The average freerunning period length for 137 human rectal temperatures was 25.00 hours ± SD 0.56 hours.

Human freerunning temperature rhythms have been claimed to *dissociate*, or freerun with different period lengths within one individual. For example, the period measured for on individual was tau = 25.7 hours for both rhythms before dissociation. After dissociation, the temperature rhythms was shorter, tau = 25.1 hours, and the activity cycle was longer, tau = 33.4 hours.[135] This was called spontaneous internal desynchronization, because there was no apparent causative event for the periods of the rhythms to depart from one another. The result implies that humans possess more than one circadian pacemaker, usually synchronized, and that they can become uncoupled from one another. In freeruns the internal phase relationships of the rhythms are not the same as during entrainment.

In comparing human rhythms with those of most other animals, we run into the problem that the favorite experimental subjects have been nocturnal rodents, whereas we humans, like birds, are diurnal. Primates, such as the squirrel monkey, are diurnal and also freerun. The period of human freeruns can be slightly altered by light intensity, opportunity to select light or temperature, exercise, or a weak electric AC field of 10 Hz.[136]

It would seem logical that the normal situation for man as well as most other organisms would be to be entrained to the prevailing environmental light–dark cycle, maintaining synchrony with its fellows and the rhythms of the natural world. However, there are reports of individual who freerun or show relative coordination, despite the presence of 24-hour Zeitgebers (light, social, temperature, mechanical clocks, etc.) in their environment.[137]

The reader should not conclude from the existence of human freeruns that it is desirable to live in a freerunning situation. The normal condition, for animals and people, is the entrained condition, synchrony with the environment and the population of one's own species.

Unusual patterns in human circadian rhythms are talked about all the time. Inability to sleep at night has been called *insomnia*. Daytime sleepiness is *hypersomnia*.

Circadian Susceptibility

There is a temporal sequence of events in the physiology of human beings. Peaks occur at different times, so that there is a temporal sequence, or a physiological schedule, of peak times. (See Figures)

Probably because of the "physiological schedule," there are also rhythms of susceptibility, for example, rhythms of sensitivity to allergens. There is a more dangerous time, early night, to experience surgery. Thus, just as there are optima in chemical and temperature conditions for living things, there are also temporal optima as a consequence of internal rhythms. The idea is that the body not only has to supply materials to the right place, but it also has to do that at the right time.

The *right time* concept has broad application in pharmacology. For example, cyclosporin is a drug from soil fungus that is used for prolonging the function of organ transplants because of its suppression of the immune system. In rats, this drug prolonged survival for an average 45 days when it was given during the light-time, but only 28 days when it was given during the dark-time.[138]

It is not unreasonable with current knowledge of the time-of-day variation in susceptibility to require that this information be obtained as part of the testing procedure for a drug. There is a problem with this idea in using nocturnal mice or rats for determination of human daily susceptibility to drugs. In practice, most drugs are administered to people during the day-time when they are awake, thus potentially imposing a 24-hour rhythm of drug concentration upon the recipient.

Some circadian scientists have envisioned a more controlled approach to administering drugs in a cyclic manner. An ingenious system, for example, was devised to administer melatonin to rats called programmed microinfusion. An Alzet Osmotic Minipump (Alza Corp., Palo Alto, CA) filled with physiological saline discharges its contents when it is implanted. The saline can be used to displace the contents of a subcutaneously implanted capillary tube precharged with a linear array of melatonin solution in alternation with melatonin-free light mineral oil. The minipump pushed fluid out of the tube and into the animal at the rate of 1 microliter per hour. In other words, the recipient received melatonin fluid for awhile, then mineral oil for awhile, then melatonin again, in a temporal program. The artificial melatonin rhythm imposed by the pump was monitored by sampling urine melatonin.[139] An implantable programmable Medtronic® pump has been used to supply a sinusoidal schedule of cyclosporine to beagles.[140]

There is also the problem that some drugs affect the underlying circadian processes themselves. For example, a study with some drugs to prevent vomiting (antiemetics) in space flight, changes were found in circadian rhythms. Dexedrine increased the daily mean heart rate and amplitude. Scopolamine reduced the amplitude of the heart rate rhythm and decreased its mean but raised the rectal temperature by 0.5°C. Scopolamine produced a phase shift in a subject.[141]

Depression and Circadian Rhythms

Brain and performance peak around the mid-day for humans. Many of the rhythms studied, for example, locomotor activity, are considered to be behavioral rhythms. In 1965, Curt Paul Richter discussed periodic physical and mental illness. His anecdotes describe cyclic psychotic attacks, a schizophrenic whose personalities occupied alternating days, periodic drinking sprees (dipsomania), and a 20-day attack cycle in a catatonic-schizophrenic.[142] Rhythms in psychiatry includes the potential relationship of rhythms to depression, and, in animals, rhythms in brain amines (norepinephrine in the hypothalamus, pineal, lateral thalamus, mesencephalon, pons, and cervical spinal cord; serotonin in the red nucleus, hypothalamus, and telencephalon), memory, and conditioned responses.

Mood disorders, or affective disorders, have periodic components. Bipolar disorder, which usually begins with depression, has episodes every 3–6 months, and rapid cycling bipolar disorder has four cycles a year or less. Unipolar disorders involve several episodes of depression over a lifetime.

The relationships of light, depression, and rhythms have attracted notice. Psychiatrists named a disorder, SAD (Seasonal Affective Disorder), in which fall-winter depression alternates with depression-free spring and summer. Winter depression symptoms are considered to be fatigue, sadness, oversleeping, overeating, carbohydrate craving, withdrawal from other people, and impaired productivity. I could argue that these "winter blues" may be a perfectly natural consequence of our evolution in a seasonal environment, in which some of the "symptoms" of SAD might be considered adaptive to a winter environment. While a winter–summer difference could be due to seasonal weather changes, investigators have focused on light and have suggested that SAD is related to the short photoperiod of winter.

Antidepressant drugs—lithium, monoamine clorgyline, and the tricyclic imipramine—slow the freerunning circadian rest–activity cycle and delay the peaks of some rhythms. Lithium lengthens the period of the rhythms in cockroaches, hamsters, desert rats, rats, and man.[143] An explanation of "affective episodes" is that some rhythms of an individual are freerunning while others remain entrained, that is that the person's rhythms are dissociated.

Table 13.1 Symptoms in More than 66% of SAD Patients.[144]

Sadness
Anxiety
Irritability
Decreased physical activity
Increased appetite
Carbohydrate craving
Increased weight
Earlier sleep onset
Later waking
Increased sleep time
Interrupted, not refreshing sleep
Daytime drowsiness
Decreased libido
Difficulties around menses
Work difficulties
Interpersonal difficulties

I first speculated about a possible role of light in 1972, at a meeting of a sleep society. Dan Kripke and I stood in a brightly lit room and discussed how melatonin was high at night, how its injection produced roosting characteristic of sleep in sparrows, how some depressed patients avoid light, and how some people seek relief from winter blues by fleeing toward the equator, where there is more light per day. This leads to a simple hypothesis, that depression is due to melatonin and, since melatonin is inhibited by light, depression might be prevented with light.

Light has been successfully used to treat SAD sufferers. Basically, the idea was to increase the winter light exposure by adding more artificial bright light (e.g. 2500 lux). Light exposure produced dramatic results, such as remissions of depression in 87 percent of patients, and it worked quickly, within a few days. If light treated was discontinued, the depression returned. The methods of phototherapy treatment varied. Pulses of light were imposed in the morning, or in the evening, or a skeleton of twice daily 3-hour pulses was given at 7:30 a.m. and 8 p.m., or an additional 5–6 hours of bright light per day.

Some manic-depressives may have disturbed circadian rhythms— reduced amplitude of some rhythms, advanced phase of some rhythms, or the occurrence of 48-hour sleep–wake cycles when they transit between mania and depression. Clinical manifestations of manic-depressive illness include early morning awakening and diurnal variation in mood. Some manic-depressives advanced their wake-up time when they came out of the depressive phase.[145]

Is melatonin involved? Manic-depressive patients were supersensitive to light (500 lux at 2–4 a.m.). The melatonin rhythm in depressed patients, like some of their other rhythms (cortisol), sometimes had lower amplitude.[146] Melatonin administration worsened depressed patients, lowered oral temperature, reduced sleep time, and affected reaction time.[147,148] However, the fact that light in the day time (when melatonin is normally low, light lowers melatonin) and the fact that atenolol (which lowers melatonin) only reduced depression in three of 19 SAD patients, has been used to argue against the *melatonin hypothesis* for SAD.[149]

While the nature of the light source used to treat depression may not matter, as long as it is bright, most investigators have used full spectrum lighting typified by the Sunbox® which provides 2500 lux at three feet using Vita-lite® tubes. Simpler measures, such as white painted rooms, bright lighting fixtures, and winter vacations in sunny locations may be effective.

The Solstice Bomber

From May 1978 to April 1995 there were sixteen mail bombings that targeted universities and airlines. The name Unabomber was given to the perpetrator of these crimes.

In May of 1995, an article in *Newsweek* (5/8/96) listed the locations and the dates of the Unabomber's attacks targeted California, and there were the most attacks (three) in 1985. One of the attacks was on June 22, 1993, a date which was significant for me, because it was the date on which I nearly drowned in a rafting accident on the Roaring Fork River.

But the distribution of the attacks was curious. There were attacks in May, May, Nov, June, Oct, May, July, May, June, Nov, Dec, Feb, Jun, Jun, Dec, Apr. Of the sixteen attacks, eight were in May and June as the summer solstice approached and passed. Four more of the attacks were in November and December, clustered around the winter solstice. There were thus 12 attacks in all as the longest and shortest days of the year approached and passed. Here was the Solstice Bomber, not a bomber for all seasons.

It was spring 1995. It was time to hear from the Unabomber again. I watched my mail anxiously. Sure enough, there was yet another attack after the *Newsweek* article, on June 28, 1995. When a Unabomber suspect was captured in April 1996, he had a bomb ready to mail. It was, after all, spring, and he was not a bomber for all seasons.

Jet Lag

When we travel east or west across time zones, we are required to resynchronize to the time at the destination. East–west travel thus requires a phase shift of the circadian biological clock. The phrase jet lag has been used to denote the symptoms.

Traveling north and south in the same time zone does not require a phase shift to adjust to a new time zone. But, there may be a change in photoperiod in north south travel, especially near the solstices.

It is impossible to separate the symptoms of circadian rhythm disruption from other possible physiological consequences of changes that occur in the environment during and because of journeys— barometric pressure, temperature, humidity, stress.

Table 13.2   Symptoms of Jet Lag.

Exhaustion
Loss of ability to concentrate
Performance decrements
Constipation or diarrhea
Insomnia
Loss of appetite
Headache
Impaired night vision
Limited peripheral vision

Recovery rate varies among individuals, and differs for the various different rhythms. One source suggests that it takes two days to two weeks to readapt to a 5–8-hour time change.[150] A crude rule of thumb has been offered that one day of adjustment is required for each 1 hour of time change.

An *asymmetry effect* has been found. One reviewer concluded that the bulk of experiments with human travelers supported a faster re-entrainment after east-to-west travel (delay) than after west-to-east travel(advance) for most travelers.

The results for travelers contrast with data for phase shifting in isolation chambers where the advancing rate was faster (70 minutes per day) than the delaying rate (54 minutes per day).[151] In this regard, people were similar to sparrows which re-entrained faster to 8-hour advances of their LD 16:8 light–dark cycles. The fasted re-entrainment occurred if the sparrows were advanced and if constant light, which made the sparrows active, intervened between the pre-shift and post-shift cycles.[152]

The explanation that has been suggested for the malaise, the symptoms of jet lag, is that different rhythms within an individual's body phase shift at different rates so that during readjustment, the person's rhythms do not have their normal phase relationships.

Prescriptions for Jet Lag

It is probable that there is a set of optimum conditions to produce the fastest possible resetting, and therefore minimum jet lag malaise, for any given trip for any one individual. But it is not so easy, in the absence of copious data on an individual, to suggest remedies applicable to everybody for all trips. Still, some attempts to do this have been made;

One simple strategy is to *remain on home time*. That is, to wake up and to retire at the usual times with respect to time in the home time zone. This strategy is feasible for short trips involving only a few time zones with a little preplanning so that meetings at the destination take place at a time that is comfortable in the home time zone. For example, traveling 3 hours west from the east coast to California, schedule meetings no earlier than 11 a.m. California time which will be 8 a.m. Eastern Time.

Another strategy is to *preadapt to the new time zone* before the trip to the new time zone. Preceding the trip, the traveler begins to rise and retire using destination time zone times. Use longer preadaptation time for longer trips across more time zones. Even partial preadaptation, three hours time change for a trip that will involve a 6-hour time change, might be helpful in reducing adaptation time at the destination.

Another strategy is to *make use of the weekend lag effect.* You can make use of this effect if you have weekend lag, and you know how much later you normally sleep on Saturday than on weekdays (the average is about 2 hours).

Travel west on Sunday. To travel west, from the east coast to the west coast of the United States, get up at the normal late weekend time on Sunday, travel west, and go to bed on west coast time on Sunday. On Monday, wake-up using west coast time. If you have a typical 2 hour weekend lag, you would have only a 1 hour adjustment to make to adapt to west coast time.[153]

Travel east on Friday. To return to the east coast, or to fly to Europe, travel on Friday, and go to sleep on east coast time on Friday. This gives you the weekend to make the readaptation to the earlier wake-up times required by traveling east (e.g. 3 hours from California to Washington DC).

Stay awake from departure until bedtime at the time of the destination. This strategy, which has the advantage of simplicity, was effective for sparrows, but might not be so easy for humans. Stay awake, and keep the lights on during the trip. Go to bed, and even if you don't sleep well, begin your day's activities at the desired time in the destination zone.

More complex programs have been outlined, and even given names, such as the Jet Lag Program. Use lighting and physical activity anticipating the day-time of the new time zone. Eat high protein breakfasts and lunches for energy and carbohydrate dinners to encourage sleepiness. Use stimulating drugs to stay awake (methylxanthines such as theophylline in tea; or caffeine and theobromine in coffee or in chocolate). This program is not simple, and it requires a feasting/famine schedule in anticipation of the trip. Its proponents claim it can reduce jet lag malaise to three days, even for a 12 hour time change.[154] A program using manipulation of the phase shifting effects of light and dark to reduce jet lag on trips longer than three days has been described.[155] The authors claim the program can reduce jet lag weeks to a few days.

I am not recommending or using a pharmacological approach myself, but there has been a great deal of interest in coming up with a pill to prevent jet lag malaise. One pharmacological strategy has been to attempt to alleviate jet lag with *melatonin*, whose exogenous administration can cause fatigue and sleep.[156] Another has been the administration of a benzodiazepine (triazolam, a drug for insomnia) to achieve phase shifts which have been gotten in animals.[157]

At the time of this writing, in my opinion, there is not a satisfactory general remedy for the malaise that has been called *jet lag*. Realizing that you may experience rapid eastward or westward travel, you can, however, certainly try to plan to allow some days of adjustment at your destination.

I travel infrequently, and my trips are often too short to make much use of jet lag reduction programs. I rather enjoy the variety provided by the opportunity to be awake at a time with respect to night and day that is unusual for me. Never seen the sunrise? Go west, you should wake up early by time in the western time zone.

Shift Work, A Circadian Problem

One form of "shift work" is the "watch" systems that are used on ships. There are a variety of these systems. For example: 4–8 a.m. morning watch, 8–12 noon forenoon watch, 12–4 p.m. afternoon watch, 4–6 p.m. the first dog watch, 6–8 p.m. the second dog watch, 8–12 a.m. the first watch, and 12–4 a.m. the mid watch.[158]

Almost 27 percent of the United States work force is employed in shift work, that is, work that is scheduled during all or part of the night.[159] For example, here is one way to divide the 24 hours into three shifts:

> 8 a.m.–4 p.m., day shift
> 4 p.m. to midnight, swing shift
> midnight–8 a.m., night shift, graveyard shift, lobster shift.

Some of the individuals on shift work schedules encounter similar problems to those of travelers. Whereas the travelers' adjustments are reinforced by the natural schedule at their destinations, the shift workers work at unusual (non-daytime) phases in an environment that provides conflicting, rather than reinforcing, social and lighting time cues. In addition, the shift workers, unlike the occasional travelers, experience repeated disruptions.

Just as for travelers, there are a number of scheduling strategies that have been adopted which may reduce the consequences of shift work.

The first line of defense is to *avoid shift work schedules* altogether.

The *fixed schedule* strategy involves hiring workers for particular shifts. This system depends on employee preferences and pay incentives. It would seem from the weekend-lag effect that it might be easiest for most workers on the 11 p.m.–7 a.m. shift to do their sleeping in a darkened room from 7 a.m.–3 p.m. An individual's entrainment to the cycle could be enhanced by the use of bright light in the work place, by adherence to a regular schedule, by use of methylxanthines, and by meal timing and content. From a circadian point of view, the fixed schedule method seems potentially healthiest, if shift work schedules must be used at all.

In the rotating shift strategy, workers take turns on different shifts. The use of one week on a shift is common practice. Circadian researchers have recommended longer times (a month) on each shift, so that there would be less frequent readjustment.[160]

Worker work schedule satisfaction, subjective health estimates, personnel turnover, and worker productivity were improved by use of *delaying schedules* and less frequent rotations (every 21 days) compared to weekly advancing rotations. This may be due to the fact that the human freerun is about 25 hours.[161]

Brighter artificial light in the work place might help. A light intensity of 3000 lux (the normal range of artificial illumination is usually less than 1000 lux) is required to synchronize human rhythms.[162]

The typical worker doing shift work is subjected to two conflicting schedules, his rotating work schedule, and the general schedule in the outside world. When we tried to mimic this with sparrows by subjecting them to LD 8:16 rotated weekly and LD 12:12 fixed at the same time, they did not follow the rotations. Instead, they responded as though they were being subject to a series of changes in photoperiods.

At the time of this writing, I don't think there is a satisfactory "remedy" for the undesirable, and possibly dangerous, physiological consequences of shift work.

Investigators who study shiftwork are fond of pointing out the times of catastrophes, especially when trying to document the importance of studying rhythms and sleep when applying for grants. Three Mile Island lost core coolant water from a stuck valve at 4–6 a.m. The Davis-Besse reactor at Oak Harbor, Ohio, lost main feedwater at 1:35 a.m. The Rancho Seco nuclear reactor near Sacramento lost DC power at 4:14 a.m. Chernobyl began at 1:23 a.m. And so on, accidents during night shifts, and the possibility that there were poor responses which might be due to sleep loss. The major zone of vulnerability is 1 a.m. to 8 a.m.[163]

How dangerous is shift work for the health of a shift worker? I can't tell you that. I will say that I have heard the rumblings of lawsuits about shift work connected illness. If I were an employer, I would be very cautious about imposing shift work schedules. Both employers and employees owe it to themselves to educate themselves about shift work and what it means for their biological clocks.

Table 13.3 Symptoms of Shift Maladaptation Syndrome.[164]

Sleep alterations
Poor sleep quality
Difficulty in falling asleep
Frequent awakenings
Insomnia
Fatigue which does not disappear after rest
Irritability
Tantrums
Malaise
Inadequate performance
Dyspepsia
Epigastric pain
Peptic ulcer
Regular use of sleeping pills

Chapter #14: The End of the Whole Mess

In case you are reading your way through this, I tried to save some interesting stuff for the end.

Feral Humans

What are the circadian rhythms of ordinary people? What do normal people, people like you and me (okay, well, I won't claim that college professors and writers or readers of this book are ordinary, or normal, people) actually do as they go about their normal lives?

The easiest thing to do was to ask some people to record their wake-up and to-sleep times for a month. We did this for 11 students in the USA and 15 students in Spain. The students all had daily cycles. They were awake an average of 16.3 and 16.5 hours, respectively, and woke up at 7:36 and 8:23 a.m. local time (respectively). Day-to-day variation was astonishing, averaging 4–6 hours, sometimes as much as 10 hours, which might be considered a 10 hour phase shift! Both groups had weekly cycles (circaseptan), sleeping later on weekends (average 2.3 and 2.5 hours), which are probably a consequence of their class schedules. The variability was the most surprising result, within individual variability and between individual variability.[165]

Wake-up and to-sleep times were recorded by four more people, older than the students, in the USA for an entire year. Their rhythms shifted when they changed time zones due to travel. Latest wake-up times were around the winter solstice; earliest wake-up times were around the fall equinox. New moon versus full moon days were not different. The one hour change imposed by standard versus daylight savings time was reflected by near one-hour changes in two of the participants. Weekend delays in wake-up time averaged 0.8–1.6 hours. Wake-up times were close to the time of sunrise, but to-sleep times were several hours past sunset, confirming the idea that we extend our active time mainly by adding awake time past dusk.[166]

Thanks to a grant from Temple University, I was able to purchase monitors (Motionlogger Actigraphs®) that could be used to measure wrist activity rhythms in people as they went about their daily lives. They were black metal boxes weighing 3 ounces, and were strapped on the non-preferred hand (in my case, the left hand) with Velcro bands. They were worn constantly, only removed for showers and swimming. I programmed the wrist monitors to collect the number of wrist movements and to report and remember them every five minutes. The data obtained were therefore a time series of counts of wrist motions per five minutes for two to four weeks.

It's no wonder scientists have a reputation for being weird. I wore the monitors myself for several years altogether, and continuously for one year. The monitor was always a conversation piece, since it looks like one of those devices sometimes strapped to prisoners. Incarcerated as I was wearing my monitor, I was able to analyze the one year record for daily, weekly, monthly, and seasonal trends. I also slept in a bedroom with uncovered windows and a skylight, which meant that I was exposed to natural dawn as well as to whatever artificial light I added before dawn or in the evening. Of necessity (my husband and I were both working), we used an alarm set for no later than 0600 every day including weekends. Alpha was 0.4 hours shorter on menstrual cycle days 8 through 18, but lunar phase had no effect. Annual changes of 1.1 hours were attributable to the changes between standard and daylight savings time.[167]

Wrist activity records from 17 college students (11–15 days each) showed distinctive individual patterns, much like finger prints. Everybody had activity in the day-time and period lengths close to 24 hours. There was individual variation in average phase—onset 5.3 hours, offset 4.5 hours, acrophase 6.1 hours. What this means, is that in the population that was studied, there can be as much as a quarter of a circadian cycle in variation between one individual and the next. Some individuals are rising while others are well past lunch-time, already in the middle of their "afternoons." All of the records showed the weekly variation with an average 2.1-hour delay in the onset of activity and 0.8 hours shorter alpha on the weekends.[168]

I was able to persuade some people who were traveling to wear the wrist monitors. The people traveled to Hawaii, California, Colorado, England, the Galapagos, Grand Cayman, Boston, and

Florida. The duration of daily activity, alpha, was 1.1 hour less at the destination, and the mean number of wrist motions was 63 wrist motions/five minutes less. The magnitude of the phase shifts differed as much as 2.3 hours from the expected changes, those required to fully adapt, to changes in time zones. In other words, you may make a trip to England, a five hour time change, but you may not shift your body clock five hours.[169]

Astronauts have complained about poor sleep, sleep loss, and fatigue during their missions. They have shifted sleep and work schedules and there is environmental noise from round-the-clock radio transmissions. Russian cosmonauts sleep when it is night in Russia, and some American crews have slept when it is night in Cape Canaveral, thus maintaining their home time schedules. Astronauts' brain recordings show they did sleep, and pilot Frank Borman had a 23.7 hour pattern when he lived on a 23.5 hour day on the Gemini VII mission.[170] Body temperature and alertness rhythms were delayed in space and had reduced amplitude. Sleep was shorter, REM latency was shorter, and slow wave sleep was shifted from the first to the second sleep cycle in the Mir space station.[171]

Table 14.1 Sample Wrist Monitor Results.

	Weekday	Weekend
Motions/5 minutes	405	387
Onset	6.9 DEQ	7.6 DEQ
Offset	23.4 DEQ	23.1 DEQ
Alpha	16.5 hours	15.5 hours
Acrophase	14.1 DEQ	13.7 DEQ

DEQ = the decimal equivalent of Eastern Daylight Savings Time. Onset is the beginning of activity, near the wake-up time. Offset is the end of activity, near the to-sleep time. Circadia software was used to analyze the data.

Yawning

Ho hum. It seems appropriate to put yawning in the last chapter.

The wrist monitors had a button that was an event marker. By pressing the button, the wearer could make a "mark" on his time record. We studied yawning by asking 6 wrist monitor wearers to press the buttons whenever they yawned. Of course, they only pressed the buttons when they were awake, so we don't know if they yawned when they slept. There was a small peak in yawning in the morning, a middle sized peak in late afternoon, and a big peak late at night. The most yawns recorded in a day was 28.

Yawning has been attributed to boredom, lack of oxygen, social contagion. In mammals yawning may be associated with transitions between periods of high and low activity or arousal. Relatively sedentary species that sleep very little, such as many herbivores, yawn infrequently if at all, whereas species that sleep 8–10 hours daily and have both activity and inactivity yawn more. Yawning has been associated with changes in hormones, testosterone and melanophore stimulating hormone. Skin conductance and heart rate change after a yawn, and in the 15 minutes following the 747 yawns, wrist motion increased reliably in all the subject's records, consistent with Maryann Baenninger's hypothesis that yawning predicts an increase in activity level. Yawning frequency is not related to amount of sleep or to an individual's personal wake-up and to-sleep times. There was more yawning during the week than on weekends.[172]

Table 14.2 Yawns and Wrist Motions in Eight People.[173]

Mean Yawns per Day	Onset Time of Day DEQ	Offset Time of Day DEQ	Alpha, Hours	Acrophase Time of Day DEQ	Motions/5 Minutes
6.9*	6.5	23.0	16.5	14.0	303
13.3	10.7	24.6	13.8	16.4	268
7.5	7.7	24.0	16.3	15.3	285
10.5*	7.6	23.9	16.3	15.2	326
1.6**	6.6	24.1	17.5	14.5	460
3.9	6.9	24.4	17.5	14.4	262
4.9**	7.4	23.6	16.2	13.9	367
8.5	9.4	26.0	16.6	17.1	311

Each line has the data in one 7–15 day recording from one individual. * two records contributed by the same psychologist; ** two records from the same biologist.

Electromagnetic Fields

We are only beginning to worry (actually, I've been very worried for a long time!) about what all the extra artificial light, light pollution if you will, in our environment might do to us and to other living things. We have mostly exploited artificial lighting to our considerable advantage.

Less obvious are concerns about the concerns of radiation other than visible light in our environment. For example, we are now exposed to artificial electromagnetic fields (EMFs) including low frequencies from power lines and higher frequency radio, television, microwave, satellite communication, and radar. The name ELF has been given to extremely low frequencies (e.g. 30–300 Hz). A continuous electric 10 Hz field affects human circadian rhythms. Period length shortens from 26.2 to 25.0 hours. The amplitude of the body temperature rhythm increases from 0.382 to 0.439°C. The mean body temperature increases from 37.054 to 37.120. The fraction of time spent in sleep decreases from 33.8% to 32.1%.[174]

Circadian Rhythms and Work in Space

Circadian, circannual, and lunar-tidal rhythms have evolved because of the daily and annual fluctuations in the lighting and environment on earth due to its rotation rate, orbiting rate, and its relationship to the sun and the moon. When we venture into space, these familiar environmental cycles are left behind.

A spacecraft orbiting the earth every 100 minutes experiences a 100-minute light–dark cycle, not a 24-hour cycle, and less than 30 percent of time is spent in the earth's shadow (dark phase).

Daylength on other planets in our solar system is not 24 hours. For example, about 15 days of sunlight alternate with 15 days of night on the moon; Mercury rotation is 59 days. Venus rotates in 243 days. Mars rotates in 24 hours 37 minutes. Saturn and Jupiter rotate in 10 hours. Uranus rotates in about 24 hours. And Pluto rotates in a little over 6 days.

During a flight, such as to the moon, without shades and artificial lights, travelers would be subject to a black sky and to constant sunshine.

The light intensity on other planets is different than on earth. On the moon the earth shine is 75 times the illuminance of the full moon.

While the traveler seems to have infinite scheduling options, in fact, his own circadian physiology, which he carries with him, limits these options.

One organism that was actually launched into space is the fungus, *Neurospora crassa*. *Neurospora* is the bread mold that makes orange bands with a circadian rhythm. Race tubes were packaged in foam and sent aboard a space shuttle. When the cultures were re-examined upon their return to earth, rhythms with a period of about 22.7 hours were observed, so the freerunning rhythm persisted in space. The cultures were affected by the spaceflight, which involved large temperature changes and other disturbances, that were visibly different from the control cultures kept on terra firma. In the space tubes, there was greater variation in the growth rates among the space tubes, greater variance in the circadian period length, and the clarity of banding, which reflects the amplitude of the circadian rhythm, was reduced.[175]

Making the Most of Time

There may be something you can gain in your own daily life by taking a less casual consideration of the role of your body watch.

For example, you could try rising at the same time every day.

And, you could increase the light to which you are exposed by less use of window coverings, brighter artificial lighting, and spending more time out of doors.

Larks and Owls

People often refer to themselves as larks (early risers) or owls (late retirers).[176, 177]

The distribution of wake-up and to-sleep times probably varies depending on factors such as occupation, and day of the week, but a sample distribution of wake-up times might be at least of interest. The data were for four Tuesdays per individual in February recorded by 35 twenty-something male and female college students residing in the Delaware Valley.

Table 14.3 Distribution of Wake-up Times.

Wake-up Time DEQ	Number of individuals
6–6.9	5
7–7.9	13
8–8.9	7
9–9.9	8
10–10.9	2

Table 14.4 Distribution of To-sleep Times.

To-sleep Time DEQ	Number of individuals
22–22.9	2
23–23.9	9
24–24.9	10
25–25.9	9
26–26.9	2
27–27.9	3

Naps

I never took naps until I was 37 years old. Then suddenly I found myself falling asleep some afternoons. I took the most naps in February.

Naps have, of course, not escaped the prying noses of scientists. Sleep duration measured with sleep diaries in 102 college students (18–22 years old) average 7.4 hours on week nights, 7.9 hours on weekend, and total sleep is 7.9 hours which includes the effects of naps. The incidence of napping is so variable among reports that it is difficult to describe with as many as 60% not napping at all.[178] Students ($N = 14$) in 1992 report on napping. Of the 14 students who gathered data for a month, 5 took some morning naps, and 7 napped in the afternoon more than 10% of days.

The range of individual average nap lengths is 0.67 to 1.50. Catnaps shorter than 15 minutes or naps longer than 2 hours were rare. There are more naps in people during the 3rd through 8th decade. There is more napping in equatorial cultures that permit a siesta period and the least napping is at the higher latitude of England. Napping is not a disease, and it can improve performance, especially in persons deprived of night time sleep. Naps longer than 30 minutes are recommended.[179]

Insomnia

Insomnia is a word used for for sleep difficulty sleeping, or disturbed sleep patterns, or insufficient sleep. Insomnia has been subdivided into five types: broken sleep, falling asleep too late, falling asleep too early, disorganized sleep–wake schedules, and non-24-hour rhythms.[180]

Patterns of insomnia are often interpretable by considering circadian rhythms.

Popular opinion holds that insomnia occurs with aging, though the idea that the elderly require less sleep is disputed. There is supposed to be a decreased amount of deep (Stages 3 and 4) sleep which begins in the 40s and 50s, and a concomitant increase in Stage 1 sleep. The ratio of time spent asleep to the total time spent in bed, or sleep efficiency, declines from 95% in adolescence to less than 75% in old age.[181]

Insomnia can be a shortening of the total amount of sleep,[182] a disappearance of stage 4 sleep, and more interrupted sleep. However natural insomnia may be, people complain of difficulty falling asleep or early morning awakening, although to me that appears to sometimes be due only to the objections of the people around them. Trouble falling asleep at night followed by difficulty awakening in the morning has been named delayed sleep phase insomnia (DSPI).[183] Sleep rhythm disturbances are associated with jet lag and shift work and if the hypothalamus is damaged.

The medical solutions to insomnia have included relaxation techniques, aspirin, and even warm milk. Drugs used are called "hypnotics" and they include the benzodiazepines (Trazolam, Temazepam, Fluazepam), antidepressants (Amitriptyline), chloral hydrate, Meprobamate, Diphenhydramine, and pyrilamine. Barbiturates are "not recommended" because of problems with dependency and a high suicide risk.[184]

Insomnia at northern latitudes, a sleep onset insomnia, a phase delay, has been treated with light.[185]

I never had insomnia either, until I was 37 years old. I rejected the plethora of sleep potions, and looked on insomnia instead as a gift, as "found time." I made use of it. Since I viewed it as a gift, it was "extra time" to read or write or use as I pleased.

It is possible that a variety of sleep patterns within the human population conferred upon us an evolutionary advantage, sort of a natural "shift work" schedule, and that while our diverse patterns might not always be comfortable for those who wake very early or retire very late, because they place one out of step with everybody else, the diversity of our sleep patterns might not be entirely unnatural.

The Future

When I spoke with publishers about this book, they advised me that it would be of most value if people could measure their sleep–wake cycles and predict the future. Indeed it would!!! I promise no such magic. About the best you can do is record your rhythm, and knowing your wake-up and to-sleep times today, use it in making tomorrow's plans.

If you have some activity, and want to optimize it, you might try measuring your performance at different times of day to see which is best for you. For example, if you run the hundred yard dash, you might be able to determine when you are the fastest dasher, or even whether time of day makes any difference at all to your running speed. With information about your best times in hand, you may be able to plan, even adjust your wake-up and to-sleep times, so that you are performing personal best time.

Biorhythm charting as a predictor would be useful, except for one thing, and that is intra-individual variability. Even within one individual, day to day variation is large. I have not promised you magic. For magic you need to turn to a fortune cookie or to gipsies who can tell your fortune with Gypsy Witch Tarot cards and read the lines in the palm of your hand. Astrology also attempts to foresee the future from the motions of the sun, and moon, and stars. None of this nonsense has anything to do with your biological clock as I have described it here.

More useful than this book would be a watch, or a watch function, or a recorder that would measure and display your rhythm. This is technologically possible. My expensive efforts to bring you a Biological Watch or a Personal Rhythm Recorder stumbled on rocky paths of obstacles and fell off cliffs of discouragement erected by a patent attorney, the US Patent Office and several watchmaking companies.

Plus Ultra

I am writing this about twenty minutes from a concrete castle named Fonthill that is in Doylestown. Fonthill is the concrete confection of Henry Chapman Mercer. He created a live-in sculpture with a book written in tiles on the floors and ceilings and walls. One of his phrases is *Plus Ultra* (More Beyond). I took this to imply the secrets in etchings, hidden cupboards, a talisman in a well that he writes about in his fiction.[186] But *plus ultra* was interpreted otherwise, and introduces the subject of immortality.

T.S. Eliot wrote that when we grow old we'll wear the bottoms of our trousers rolled. These trousers will be white flannel. Am I saved from old age by wearing a rayon skirt? He also asks if we will dare to part our hair as we please, hear singing mermaids, walk on beaches, and eat peaches as we choose. He was an optimist. I can tell you that what you can do will depend on the rules imposed on you by your old folks home. On my last visit to such a place I saw no beaches or peaches, no one was wearing trousers, few were walking, and I heard only the voices of television sets.

Just as I cannot tell your fortune, I cannot offer you a drink from the Fountain of Youth. CUPID is the inapt mnemonic for cumulative university progressive intrinsic and deleterious changes which we call aging. Most physiological variables decline with age. Each species has a characteristic lifespan, and that has been referred to as a biological aging clock or the *longevity clock*. Baboons live at most 27 years, cats 28, dogs 20, brown bears 36, horses 46, elephants 70, house mice 3, gray squirrels 15, Herring gulls 41, Galapagos tortoise over 10, toads 36, humans 115, and guppies 6.[187]

The pineal glands of humans exhibit an increase in calcification with age. The calcium deposits, or acervuli, are visible to the naked eye and are called "brain sand." I've seen pineal sand myself, and it is cloudy translucent yellow. The largest chunk I saw was over a sixteenth of an inch in diameter. This discovery prompted Fred Mustard Stewart to write a novel called *The Methuselah Enzyme* in which an enzyme named Mentase secreted by the pineal gland was responsible for aging.[188] Fiction, gang, it was fiction.

The idea that the pineal gland might be involved with aging did not die with fiction. When 575-day-old male mice were given 10 micrograms/milliliter of the pineal hormone, melatonin, in their drinking water from 6 p.m. to 8:30 a.m. each day, they lived to 931 days. Control mice only lived 755 days. Melatonin, made in the darktime by retina and pineal, is a small molecule in a family named indoleamines that includes serotonin and a plant hormone named auxin. The work with mice has led to the proposal to consume 0.5–5 mg of melatonin (depending on your age) at bedtime for "age reversal."[189]

When I walked into the City Market and found "The Miracle of Melatonin," by James O'Brien, Globe Mini Magazine #1634, for sale at the checkout counter, I knew my hormone had "arrived." Large red signs in my local health food store yelped: "We have Melatonin!!!"

Melatonin is not a prescription drug, and is being sold over the counter in pharmacies and health food stores. I am not taking it myself, and I worry that perhaps I should caution you about the possibility of possible adverse reactions. Melatonin from Silastic capsules makes hamster testes get small and melatonin in capsules or drinking water makes birds lose their circadian freeruns. We have the idea that melatonin reduction by light (which might increase the availability of its substrate serotonin) is a way to treat depression, and that leads to the idea that melatonin is depressive. The dose of melatonin being recommended for "age reversal" is small. 2.5 milligrams of melatonin injected into house sparrows caused them to assume roosting posture and lowered their body temperatures, both characteristic of their rest.

CAUTION!!!

I am not the only one "going slow" on popping melatonin pills. Long time melatonin experts Alfred Lewey (Oregon Health Sciences University, Portland), Charles Czeisler (Brigham and Women's

Hospital, Boston), and Richard Wurtman (Massachusetts Institute of Technology, Cambridge) are all suggesting that scientific evidence does not yet support use of melatonin to treat jet lag, or insomnia, or as a fountain of youth.[190]

I have found it quite curious after a quarter of a century of my career to find that, in the course of studying the effects of light and dark on the timing of circadian rhythms, I have devoted my studies to a gland—the pineal gland—which Descartes, in an inspired sophistry, suggested was suspended above a duct in which it moved the animal spirits in the anterior cavities (ventricles) of the brain so that they could communicate with those of the posterior in its function as...the soul. The soul![191] And, if function as the soul were not enough, it is now suggested that melatonin, the pineal hormone, might be the very elixir spraying from the fountain of youth.

In April and May, 1993, I counted the stones of Stonehenge. That's supposed to be bad luck, but I was never very superstitious. Perhaps I should have been more careful about the old stones.

June 22, 1993 I was in a raft whose guide flipped its seven occupants into the 44°F raging rapids of the Roaring Fork River. Before my rescue by the heroic Rich Zelter, I had a near death experience. No amount of melatonin would have saved me from that river.[192] Pineal and melatonin aside, I saw no More Beyond, only myself, a small yellow jacketed speck tossed like a woodchip in the brown water hundreds of feet below. I never thought I was particularly important, and the experience confirmed that, but, I would echo the opinion offered by Mr. James Michener after his first airplane crash (he was in three) that the realization of your own unimportance sets you free.

I also know what its like to emerge from hibernation, since I had hypothermia (body temperature 88.7°F, about ten degrees below normal). I was rewarmed under a piece of plastic with circulating hot water in a hospital. It's a shivering experience that feels real good.

After Georgia O'Keefe died, my colored pencil drawings of flowers took on a new richness, and I wondered if Georgia's spirit was in the air, musing artists here and there. And then I had a dream in which I was allowed to ask the great Oz about death and the afterlife. The Voice of Oz (I couldn't see him) informed me that our "spirits" flicker a few minutes or at most an hour, somewhat like a candle flame before they gutter out and are gone forever. So you can see, I have had to abandon science altogether and wax imaginative in order to comment on the matter of immortality.

In 1990, I read an article in which Michelangelo's *Creation of Adam* was interpreted in which a way that the composition of the fresco represented neuroanatomy. A dangling angel's leg is in the location of the pituitary, God and his surrounding angels are in the shape of the human brain, the billowing red drapery is the cranium. I thought that was amazing, five hundred years having gone by and Lynn Meshberger had captured the spirit of Michelangelo by appreciating the painting in a totally new way.

Looking more closely at Michelangelo's work in my books, I noticed that his *Last Judgement* is composed in the manner of a huge pumpkin face. The vaulting of the chapel shapes the eyes. The eyes, nose, and mouth regions are lapis lazuli blue, as if one were inside the skull looking out at the blue sky. If one imagines the Sistine Chapel as a sculpture of a skull, and I think that Michelangelo the sculptor probably did just that, then what he painted is the pineal "eye" view—people naked as jaybirds trumpeting in the hole that would be the mouth. Michelangelo himself dangles like a pip of mucus from the nasal region.

T.S. Eliot wrote that women will come and go in a room and speak of Michelangelo. That prediction I could make come true, and have done. If you want immortality, build a castle, write a book, paint a chapel.

So you see, when it comes to the subject of immortality, I have had to abandon science for the moment, and resort instead to poetry and art, and still, the best I offer, is say to you that we are all out of time. I wish you ... contentment.

FIGURES

Figures

The figures have been gathered into one section for easy comparison.

From Sparrow Perch Hopping Experiments

1. Reading A Circadian Record
2. Entrainment
3. Skeleton Cycles
4. Freeruns
5. Phase Response Curve

From Human Wrist Activity Experiments

6. Daily Cycles
7. Bargram
8. Average Wave Form
9. Periodogram
10. EA Gram
11. Alarm Clocks
12. Travel to England
13. Travel to Hawaii, Average Wave Forms
14. Annual Rhythm of Activity Onsets and Ends

From Human Wake-up and To-sleep Records

15. Circadian Rhythm Time Chart
16. Traveler
17. Weekend Delay
18. Annual Rhythm of Wake-ups and To-sleeps

Other

19. The Menstrual Cycle
20. The Physiological Schedule

Inventions

21. The Biological Watch
22. The Rhythm Recorder

Figure 1   Reading A Circadian Record

We have body temperature rhythms of about one degree, but sparrows have a much larger amplitude daily body temperature rhythm, sometimes as much as 5 degrees. The record shows the circadian rhythm of sparrow body temperature that was recorded using a telemeter that was implanted in the sparrow. The record illustrates entrainment in LD, freerun in DD, and the arrhythmia in LL.

SPARROW BODY TEMPERATURE RHYTHM
ENTRAINMENT, FREERUN, AND ARRHYHMIA

At A, a sparrow was placed in LD12:12, 12 hours of light in alternation with 12 hours of dark. It's body temperature entrained. The body temperature was high during the light-time.

At B, the bird was placed in DD, constant dark. It's body temperature rhythm freeran with a period greater than 24 hours which accounts for the "drift" of the pattern to the right.

At C, the bird was placed in LL, constant light. It's body temperature rhythm disappeared.

The bar represents the LD12:12 light-dark cycle to which a sparrow was exposed during A.

DAYS

24 HOURS

The diagram shows the record for a single house sparrow, caged individually in a soundproof chamber, and treated first with an LD12:12 light-dark cycle, then with constant dark (DD), and finally with constant light (LL). Each line of data represents 24 hours of body temperature recording (high body temperature during the light of LD12:12). The lines have been arranged vertically in chronological order, so that time reads like print in a book with the earliest data at the upper left and the last data at the lower right.

Figure 2   Entrainment

The Figure is a raster plot of an event record of sparrow locomotor perch-hopping activity showing the effect of photoperiod on entrainment. Each line is 24 hours of data. The ends of each line are midnight. Noon is in the middle. The lines are arranged vertically in chronological order. So the *x*-axis is 24 hours, and the *y*-axis is time in days.

A. LD 12:12. The sparrow was active in the light. The lights were on 6 a.m. to 6 p.m. B. LL. In dim constant light (240 lux), the sparrow "freeran" with a period shorter than 24 hours. C. LD 1:23. The period of the rhythm shortened, and the activity drifted to the right as the sparrow entrained. The heavy burst of activity is during the light and illustrates the direct effect of light, masking. D–N. The duration of light was gradually increased in 2-hour steps from 2 to 22 hours. The effect of light duration, or photoperiod, is shown, as increased photoperiod lengthened the time that the sparrow was active. O. LD 23:1. P. LL (870 lux). The bright light made the sparrow arrhythmic.

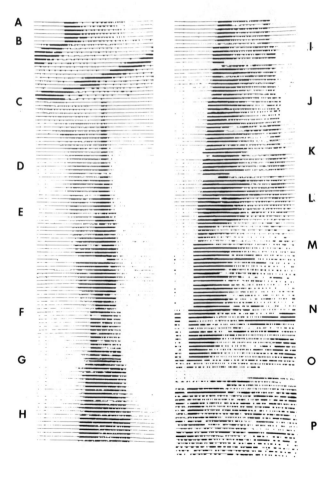

Figure 3　Skeleton Cycles

The figure is a raster plot of an event record of sparrow locomotor perch-hopping activity showing the effect of a skeleton cycle. Each line is 24 hours of data. The ends of each line are midnight, noon is in the middle. The lines are arranged vertically in chronological order. So the *x*-axis is 24 hours. The *y*-axis is time in days.

The sparrow was alternately exposed to LD 12:12, lights-on 6 a.m. (A, C, E, G, I, K) and LD 1:11 (B, D, F, H, J, L). The lights-on time of LD 1:11 was varied experimentally which phase shifted the sparrow's rhythm. The sparrow entrained to all the cycles. In LD 1:11 there was heavy activity where the light was on (direct effect, masking). Transients occurred when the bird reentrained (B, D, G, H).

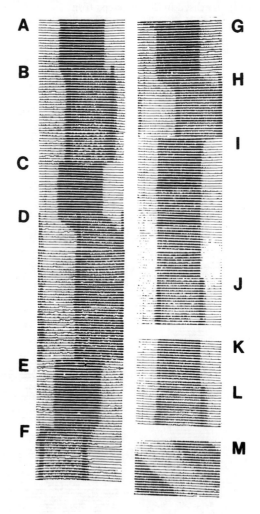

Figure 4   Freeruns

The figure is a raster plot of an event record of sparrow locomotor perch-hopping activity showing the effect of light on the freerunning period. Each line is 24 hours of data. The ends of each line are midnight, noon is in the middle. The lines are arranged vertically in chronological order. So the x-axis is 24 hours, and the y-axis is time in days.

DD, upper record. The sparrow's freerunning period was longer than 24 hours.

LL, lower record. The sparrow's freerunning period was shorter than 24 hours. It's activity time was lengthened by the LL.

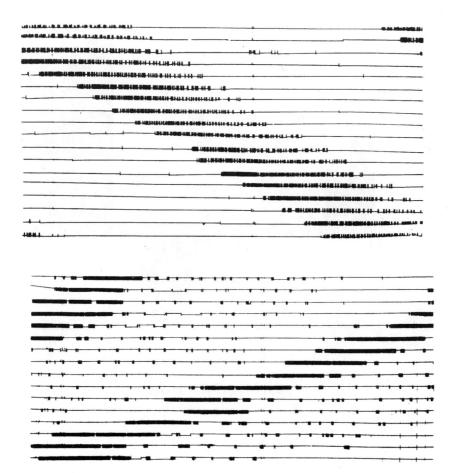

Figure 5   Phase Response Curve

The graph shows a phase response curve for sparrows. Each bar represents the average shifts of a group of sparrows exposed to one light pulse given at the circadian times below. The tiny "tees" represent one standard error of the mean, a measure of variation in the data. Circadian time zero was the phase of the activity onset prior to the experiment with the 4-hour light pulse. The sparrows normally begin activity close to the time of lights-on of LD 12:12. There is a relatively "dead zone" or region of insensitivity at 4, 6, 8. There is a "singularity" at c.t. 18 when the sparrows changed from delays to advances (Xing).

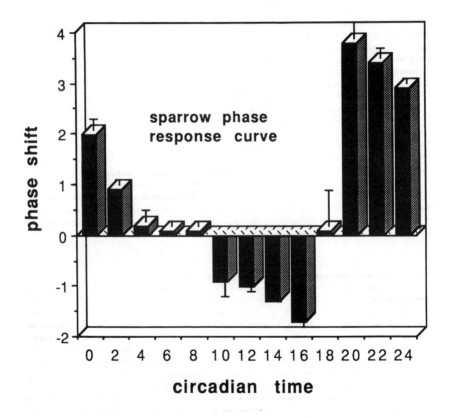

Figure 6   Daily Cycles

The graphs are four barstrips of wrist motions from one individual human. The graphs were made with Circadian Software. The *y*-axis is activity=wrist motions/ 5 minutes. The *x*-axis is Eastern Daylight Savings Time beginning on the left with 0000 hours (midnight).

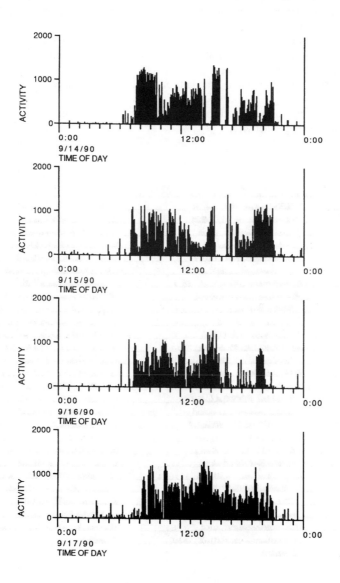

Figure 7　Bargram

The graph is a double plotted bargram. It is from the same individual woman, living at home, and working a weekly schedule. The bargram was made with Circadia Software. The x-axis (48 hours) is time beginning with 0=midnight on the left. The y-axis is time in days.

The data are double plotted horizontally so that the entire activity and rest can be seen with 0=midnight on the left. The space between bars (the y-axis for one bar, is 1500 wrist motions).

Bargram plotting displays like this one show the rhythmic qualities of the daily activity cycle. It shows the variability in the phase of the beginnings and ends of activity, and variation in amounts of activity.

Dates are printed on the left. The daily activity average motions/five minutes are printed on the right.

The method measures wrist motions, not sleep, but the individual was usually asleep in bed during inactivity.

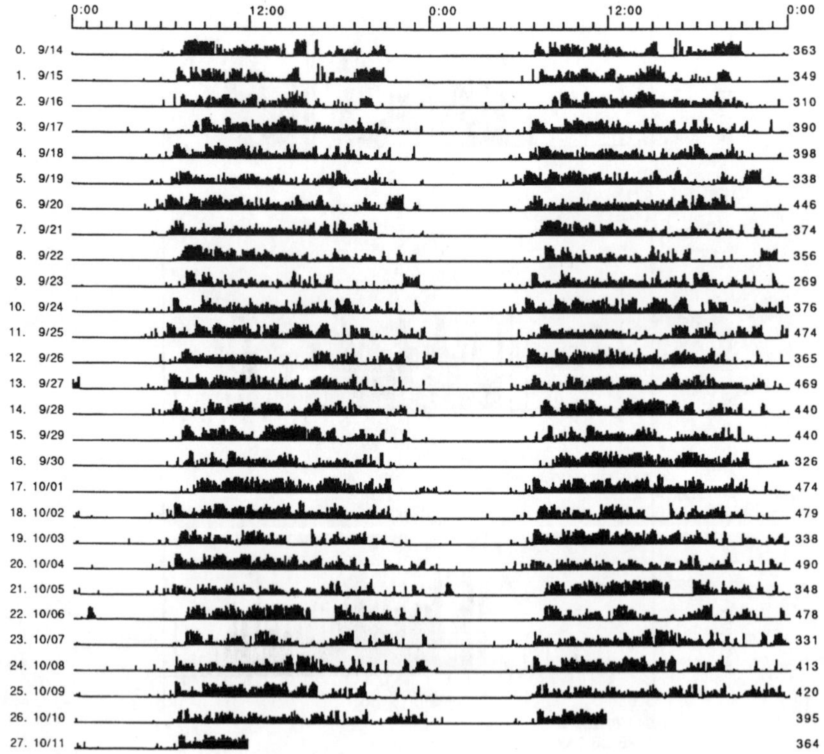

Figure 8   Average Wave Form

The average wave form graph for the same data from one person as Figure 7 is shown. Circadia software was used to make the graph. The vertical y-axis is average wrist motions per five minutes. The horizontal axis is time in hours.

The large early day activity peak corresponds with the person's self-assessment as a "morning" person.

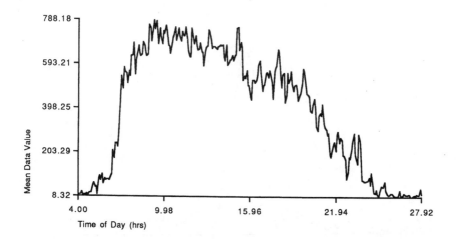

Figure 9   Periodogram

The periodogram is a method of time series analysis for detecting the presence of a rhythm and estimating the period length of rhythms.

The graph is a periodogram calculated using Circadia software for the same data from one person as Figure 7. The peak value of the periodogram represents the period length of the rhythm, 24.0 hours.

The Q value, y-axis, is a representation of the amount of "similarity" between a data record and the average waveform by comparing the standard deviations of the data with the average waveform. The x-axis is period length in hours. The method is that of Dorrscheift and Beck (*J. Math. Biol.* 2, 1975, 107–121).

There are other methods of periodogram calculation.

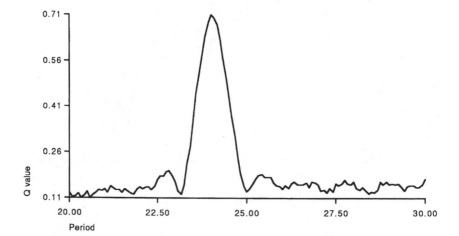

Figure 10   EA Gram

An EA Gram is an "event" record (similar to the event records for sparrow perch-hopping activity). Figure 10 shows the same data from one person as Figure 7 as an EA Gram plotted using Circadia.

Onsets of activity are represented with diamonds. Acrophases are marked with circles. Offsets, ends of activity, are denoted with triangles.

The onsets and offsets were calculated using an arbitrary threshold of 500 wrist motions/five minutes.

The *y*-axis for each line is thus "events" (activity or not activity per 5-minute interval). The *x*-axis is time (24 hours) plotted beginning at 0400 hours (0400 Eastern Daylight Savings Time) so that the activity period is not interrupted by midnight.

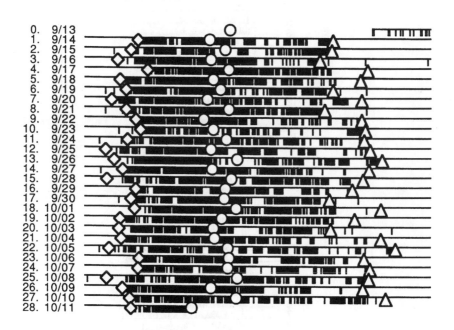

Figure 11   Alarm Clocks

Bargrams of two different individual humans are shown with vertical lines drawn to show the average time of their alarm settings.

The vertical *y*-axes are time in days; the horizontal *x*-axes are time in hours.

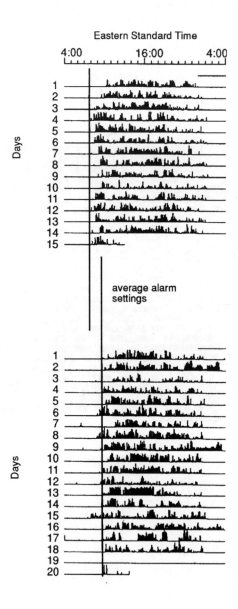

Figure 12 Travel to England

The record shows an EA gram of wrist motion of one person. The horizontal *x*-axis is time in hours. The vertical *y*-axis is time in days.

The person's circadian rhythm advanced during travel to England (6/8) from the Delaware Valley (6/16) Eastern Pennsylvania, USA, and delayed upon return to the Delaware Valley. The travel involved a five hour time change.

The horizontal axis is time Eastern Daylight Savings Time. The numbers on the right represent average wrist motions/five minutes per day. The circles are acrophases calculated with Circadia. The lines were fitted to the acrophases before, during, and after the travel.

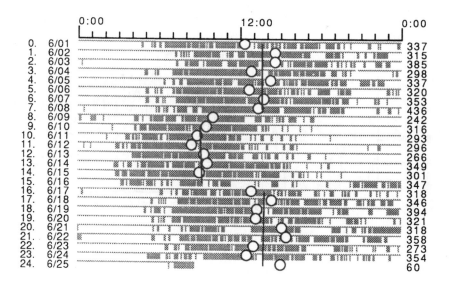

Figure 13   Travel to Hawaii

Average wave forms were calculated for wrist activity records from an individual while they were at home in the Delaware Valley and again when the same individual travelled in Hawaii.

The *x*-axis is time (adjusted to Eastern Daylight Savings time) and the *y*-axis is average wrist motions per five minutes.

The average wave forms show the adjustment to the time change (a 5 hour delay) and less wrist activity (lower amplitude) in Hawaii than at home.

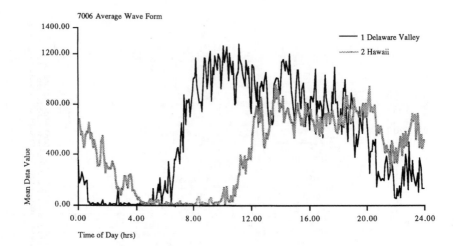

Figure 14   Annual Rhythm of Activity Onsets and Ends

The times of sunrises, sunsets, onsets, offsets, and acrophases ($n =$ one determination per day) for one individual person are graphed on the y-axis versus Julian Date of the year (x-axis).

A slight curvature visible in the graphs of onsets and acrophases can be explained by the phast shifts that the person made adapting to changes between Standard and Daylight Savings Time.

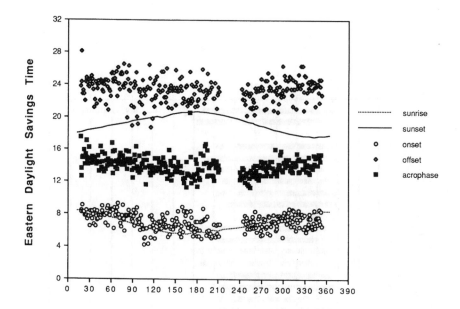

Figure 15   Circadian Rhythm Time Chart

The graph of the circadian pattern of the author made by writing down wake-up and to-sleep times and using the method in this book. The author was living in the Delaware Valley and had a weekly work schedule during this record. For publication, the graph was plotted with Cricketgraph Software.

The horizontal *x*-axis is time in hours. The vertical *y*-axis is time in days. The daily horizontal "bars" show awake time.

The time chart is a sample of a time chart made using the method in Section III of this book.

Circadian Rhythm Time Chart
for one subject for a month

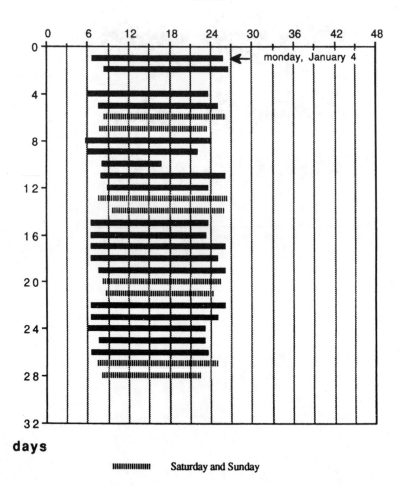

time

days

⠿⠿⠿⠿⠿⠿   Saturday and Sunday

Figure 16   Japan Travel

This is a graph of the record of wake-up and to-sleeps of a person at home in the Delaware Valley of Pennsylvania and travelling abroad, in Asia. The graph was made using the method of recording wake-up and to-sleep times in this book.

It is duplicated horizontally (double Plotted) so that the horizontal axis is 48 hours. The vertical axis is time in days. Each line then represents 48 hours of data, beginning on the left at midnight. It is graphed using Eastern Daylight Savings Time throughout so that the phase shift caused by the travel to Japan is visible.

The individual was at home in the Delaware Valley at A, flew to Japan B, and Hong Kong C, and returned to Eastern Pennsylvania D.

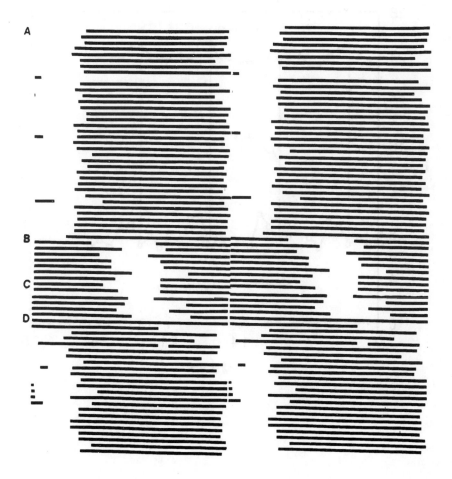

Figure 17   Weekend Delay

The graph shows the average wake-up and to-sleeps recorded over a month by eleven people.

The vertical *y*-axis is Eastern Daylight Savings Time. The horizontal *x*-axis is days.

The graph illustrates the recurrence of a weekly cycle of delays in wake-up and to-sleep times corresponding to the weekly work week. In the figures, vertical dashed lines connect the wake-up and to-sleep times on Fridays.

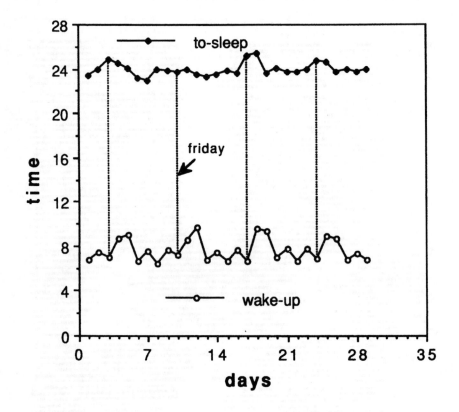

Figure 18   Annual Rhythm of Wake-ups and To-sleeps

The figure shows graphs of one individual's monthly average wake-up time (DEQ) and to-sleep time (DEQ) and awake time (hours) throughout the year.

The vertical *y*-axis is Eastern Daylight Savings Time. The horizontal axis is month of the year (1 and 13 = January)

Eastern Daylight Savings Time was used by the person from April to October, and the average wake-up and to-sleep times were thus lower in those months than in the winter months on Eastern Standard Time. Data collected while the person was traveling in other time zones were excluded.

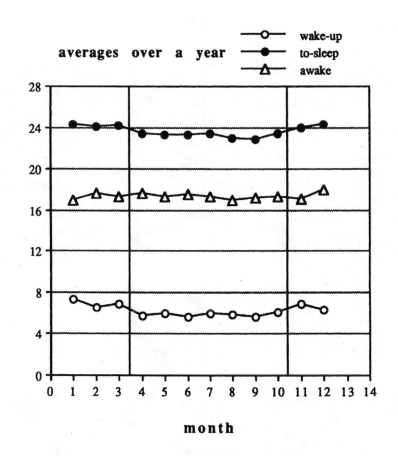

Figure 19   The Menstrual Cycle

Menstrual cycles are graphed for five women using a raster method.

Here the horizontal x-axis is 28 days. The vertical y-axis is time in years. The black bars represent five days of menstruation.

The graphs illustrate the rhythms and the individual variability of the human menstrual cycle. The cycle of the woman in b was usually shorter than days. The cycle of the woman in a was usually longer than 28 days.

c presents the beginning of a regimen of birth control pills in record (f).

P represents pregnancies in record (d).

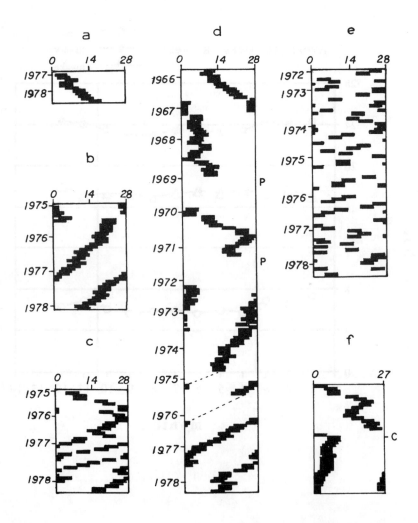

Figure 20   The Physiological Schedule

The graphs represent the sequence of events in human measurements.

The horizontal *x*-axis is time. The "dots" represent acrophases, the horizontal bars represent variation among individuals. The physiological events are arranged vertically by the time of their peaks so that those at the top of the charts occur early in the morning and those at the bottom are at night.

The large chart shows "normal" physiology within an individual. The small chart shows items related to adversity.

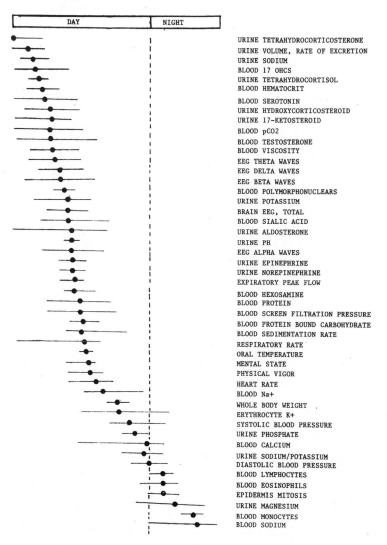

DAY	NIGHT

URINE TETRAHYDROCORTICOSTERONE
URINE VOLUME, RATE OF EXCRETION
URINE SODIUM
BLOOD 17 OHCS
URINE TETRAHYDROCORTISOL
BLOOD HEMATOCRIT

BLOOD SEROTONIN
URINE HYDROXYCORTICOSTEROID
URINE 17-KETOSTEROID

BLOOD pCO2

BLOOD TESTOSTERONE
BLOOD VISCOSITY

EEG THETA WAVES
EEG DELTA WAVES

EEG BETA WAVES
BLOOD POLYMORPHONUCLEARS
URINE POTASSIUM

BRAIN EEG, TOTAL
BLOOD SIALIC ACID
URINE ALDOSTERONE

URINE PH
EEG ALPHA WAVES
URINE EPINEPHRINE
URINE NOREPINEPHRINE
EXPIRATORY PEAK FLOW

BLOOD HEXOSAMINE
BLOOD PROTEIN
BLOOD SCREEN FILTRATION PRESSURE
BLOOD PROTEIN BOUND CARBOHYDRATE
BLOOD SEDIMENTATION RATE
RESPIRATORY RATE
ORAL TEMPERATURE
MENTAL STATE
PHYSICAL VIGOR
HEART RATE
BLOOD Na+

WHOLE BODY WEIGHT
ERYTHROCYTE K+
SYSTOLIC BLOOD PRESSURE
URINE PHOSPHATE
BLOOD CALCIUM

URINE SODIUM/POTASSIUM
DIASTOLIC BLOOD PRESSURE
BLOOD LYMPHOCYTES

BLOOD EOSINOPHILS
EPIDERMIS MITOSIS

URINE MAGNESIUM
BLOOD MONOCYTES
BLOOD SODIUM

Figure 20   *(Continued)*

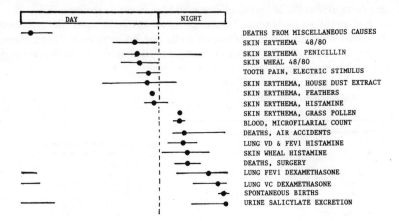

Figure 21   The Biological Watch

The Biological Watch displays the circadian rhythm of wrist activity (an actual month long record of wrist activity is displayed) for one person, its user.

Each line of data represents 24 hours of rest and activity (an actual month long record is displayed). Each line of data represents 24 hours of rest and activity events (the black regions represent activity). "Today's data" is the bottom line (only partially completed). The current body time (at the bottom) can be compared with the record (above). When the display area is filled, the record across so that the bottom line is the current day.

Inventor: Sue Binkley Tatem

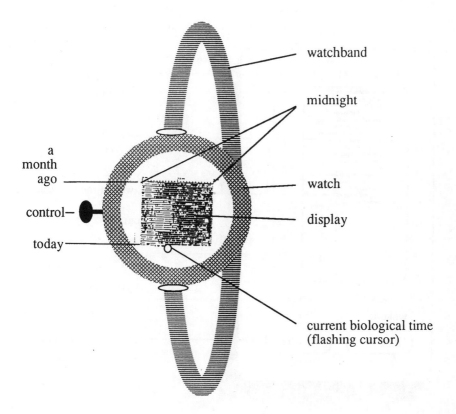

Figure 22   The Rhythm Recorder

The Rhythm Recorder displays 17 days of the sleep wake cycle for one person, its user. Each line is a day of data. Today's data is on the bottom. Yesterday is the line above. Successive lines show days that were recorded as the days ensued. When the recording reaches the bottom (28 days, 4 weeks) the record scrolls so that the bottom line is always today's record.

When the user wakes up in the morning, he presses "wake-up." The recorder begins to display a bar ending at the current time in the home time zone. When the user goes to bed he presses to-sleep and the bar ends at that time. The device has internal clock and calendar so that the wake-up times are automatically placed in the correct locations on the display.

Sue Binkley Tatem, Inventor

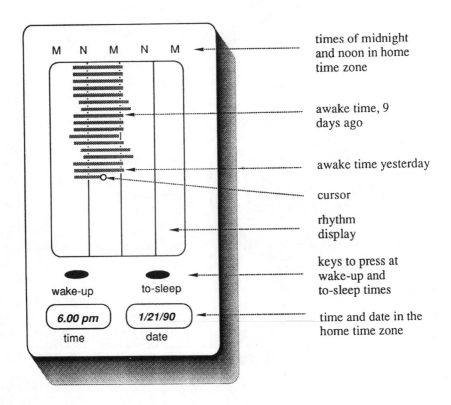

times of midnight and noon in home time zone

awake time, 9 days ago

awake time yesterday

cursor

rhythm display

keys to press at wake-up and to-sleep times

time and date in the home time zone

SECTION II: APPENDICES

Glossary of Terms and Abbreviations

The definitions provided here are for terminology used in this book and in the field of circadian rhythms. More may be found by consulting the index or a dictionary

absolute time a system intended to indicate the same instant by the same standard in all parts of the world, irrespective of local standards and without reference to the meridian under which an event takes place

astronomical time mean solar time reckoned from noon through the twenty-four hours

a.m. am, A.M., AM, before noon, used to designate the time from midnight to noon

acrophase phase angle of the maximal value of a sine function fitted to the raw data of a rhythm; a measure of the time of the peak of a rhythm

ad libitum free choice of feeding or light in experiments

advancing phase shift phase shift where one or more period is shortened; conventionally given a positive ($+$) sign

aftereffects characteristics of a rhythm which derive from pretreatment conditions, such as transients in phase or changes in period length

alpha activity time, the duration of activity

amplitude (of a rhythm) maximum minus minimum value, excursion

antigonadal inhibitory to the reproductive system, causes the testes and or the ovaries to decrease in size or weight or function (regress) or to be come atrophic

arrhythmic lacking a discernible rhythm; sometimes results from pinealectomy or constant light

asynchrony lack of synchronization between two rhythms

beta receptor theoretical membrane receptor for catecholamines (e.g. norepinephrine) that is defined pharmacologically (based on responses to stimulating and blocking agents). A beta receptor has been proposed for the control of pineal melatonin production

biological clock mechanism within an organism that is capable of generating repeated cycles (oscillations, rhythms), whose period is relatively insensitive to temperature, and which can be synchronized by environmental stimuli

cAMP cyclic adenosine 3', 5' monophosphate, cyclic AMP

circadian rhythm a rhythm with a period length of about 24 hours

clock an instrument for reckoning time using a mechanical mechanism of springs, weights, balance wheels, pendulums, or electronics that indicates time (hours, minutes, seconds) on a plate with moving hands or in a display

crepuscular active at twilight or just before dawn

critical photoperiod the duration of light in excess of which a photoperiodic response is that of a "long" day

c.t., Ct, ct, CT; circadian time, time from the organism's point of view in which the organisms circadian cycle is set "equal" to 24 and each 1/24 of the cycle is considered to be an "hour." For diurnal animals, c.t. 0 by convention = onset of activity, and for nocturnal animals, c.t. 12 by convention = onset of activity

cycle a period of time during which a sequence of events occurs—one such period is a cycle, but the use of the word implies that the sequence of events repeats

D dark-time, usually of a light–dark cycle

DD constant dark

dark-time night in natural lighting; the time during which the lights are turned off in experiments

day the light period of time between dawn and dusk, or, alternatively, 24 hours

daylight-saving time 6:00 A.M. Daylight Saving Time (DST) is the same as 5:00 A.M. standard; *daylight saving*; a method of saving or making use of daylight by setting the clock ahead, usually one hour, of standard time: this gives an hour more of daylight at the end of the usual working day

delaying phase shift phase shift resulting from lengthening of one or a few periods; conventionally designated with a minus sign

desynchronization two rhythms with separate periods

diapause period of delayed growth or development, inactivity, and reduced metabolism (as in fruit flies)

diurnal active in the day-time

endocrine gland a ductless gland which secretes chemicals called hormones into the blood (as the pineal gland)

entrainment synchronization of a rhythm in the absence of a synchronizing signal; the period of the freerun, tau, is considered to be the innate period of the rhythm (=free run)

frq frequency, a *Neurospora* gene

frequency reciprocal of period; to determine the frequency, divide one by the period length (for example, if a period = 24.0 hours, then frequency = 1/24 = 0.04166 cycles/day)

g gram

h hour, hr., hr, 60 minutes

5-HT 5-hydroxytryptamine; serotonin; precursor of melatonin

hormone chemical product of a ductless gland carried to its target via the blood; less rigorously used to denote any chemical messenger

interval timer a nonrepetitive timer such as an hourglass

in vitro outside the animal, as in organ culture, cell culture, or superfusion culture (as a pineal gland in a Petri dish)

in vivo in the living animal

jet lag the syndrome of symptoms that is believed to occur when a person's physiological rhythms are out of phase with the environmental light–dark cycle and each other (e.g., after transmeridian flights) or during shift work schedules

L light-time

LD 14:10 light–dark cycle consisting of 14 hours of light in alternation with 10 hours of dark; sometimes designated alternatively as 14L:10D

LL constant light

lunar day the time between two successive transits of the moon on a given meridian

m meter, or minute

M midnight; or, in context, M phase of cell division

melanophore pigment-containing cell in the skin

melatonin M, MT, MEL; methoxyindole hormone of the pineal gland and retina

min minute, min., 60 seconds, 1/60 hour

molecular biology deals with the large molecules of living organisms, nucleic acids (RNA and DNA) and protein

N used in context, N = noon when time is referred to, N is also used to indicate number of samples in experiments

NAT N-acetyltransferase activity; NAT is a pineal enzyme which converts serotonin to N-acetylserotonin; noteworthy for its large circadian rhythm which is controlled by light and dark

NE norepinephrine, a catecholamine that stimulates melatonin synthesis in some species

night the darkness that occurs between dusk and dawn

nocturnal active in the night-time

nychthemeron an entire day, 24 hours

oscillation a cycle

pacemaker driving oscillation, a rhythm that controls other rhythms *per period*, a *Drosophila* gene that controls production of the protein (PER) which involved in control of period length

period the length of a cycle; designated by the Greek letter tau

phase the instantaneous state of a rhythm, designated by the Greek letter phi

phase angle the difference in time between the phase of an event and the phase of another event (e.g., the difference in time between activity onset and lights-on)

phase reference point event used to calculate phase, required for determining phase angles

phase response curve plot of phase shifts in response to pulses plotted versus time the pulse was given

phase shift a displacement of a rhythm along the time axis, an advance or a delay, designated by the Greek letters, Delta phi

photoperiod light-time, length of the light-time, daylength

photoperiodic phenomenon that is responsive to daylength (or nightlength)

photoreceptor a structure that detects light

p.m. pm, P.M., PM, afternoon; the time from noon to midnight

PRC phase response curve

Q10 the temperature coefficient = rate at temperature $(X + 10°F)$/rate at temperature X

range of entrainment range of Zeitgeber periods with which a rhythm can synchronize

rapid plummet drop in N-acetyltransferase or melatonin that occurs in response to light at night with a halving time of five minutes or less

recrudescence (e.g., of testes) growth, increase in size, renewal of function; normal part of a seasonal cycle

refractory period portion of time including the late subjective night and early subjective day when melatonin synthesis (N-acetyltransferase) cannot be stimulated by placing the organism in the dark

regression (e.g., of testes) involution, reduction in size, diminution of function; normal part of a seasonal cycle

resonance experiment a protocol in which groups are subjected to cycles whose periods are multiples of 24 hours (24 hours, 48 hours, 72 hours), and compared to groups subjected to cycles with periods that are non-multiples of 24 hours (36 hours, 60 hours)

retinohypothalamic tract RHT; nerve tract from the eye to the hypothalamus

rho rest-time, the duration of rest

rhythm a pattern of events that recurs; a series of cycles

RIA radioimmunoassay; a method of measuring a hormone or other molecule that involves binding characteristics of antibodies

SCG superior cervical ganglia (or ganglion), in the neural pathway for control of the pineal gland

SCN suprachiasmatic nucleus (or nuclei), a region of the hypothalamus, bilaterally paired, believed to be the location of the circadian pacemaker in mammals

SD standard deviation, a statistic which is a measure of variation in data

SEM standard error of the mean, a statistic which is a measure of variation in data

serotonin 5-HT, a precursor for melatonin

sidereal time time expressed with reference to the stars, about 4 minutes shorter than 24.0 hours

Silastic capsules silicon tubes with plugged ends filled with a substance (e.g. melatonin crystals); a means of administering a hormone continuously

singularity (T*S*) a light pulse of specified intensity and duration which annihilates the rhythm of fruit flies

skeleton photoperiod a light–dark cycle in which dawn and dusk are represented by two separate light pulses

solar time time measured with reference to the earth's motion in relation to the sun

standard time the official civil time for any given region; mean solar time, determined by distance east or west of Greenwich, England: the earth is divided into twenty-four time zones extending from pole to pole

subjective day the time during which the organisms exhibits the behavior usually associated with light in an LD cycle; for example, a diurnal sparrow is usually active in its subjective day

subjective night the time during which the organism exhibits the behavior usually associated with dark in an LD cycle, for example, a sparrow usually rests in its subjective night

subsensitivity reduced ability of melatonin synthesis to response to stimulation, e.g. by isoproterenol; occurs after prolonged stimulation

superior cervical ganglia SCG; a pair of sympathetic ganglia located in the neck; relays neural signals to the pineal gland

supersensitivity increased ability of melatonin synthesis to respond to stimulation, e.g. by isoproterenol; occurs after stimulation deprivation, e.g., by denervation

suprachiasmatic nuclei SCN; (nucleus) region of the hypothalamus, thought to generate circadian rhythm information that is conveyed neurally to the pineal gland in some species (e.g. rats); bilateral

T period of a Zeitgeber cycle

target cells, tissues, or organs upon which a hormone acts

tau period length of a circadian rhythm

T experiments experiments in which the period of the Zeitgeber cycle is varied experimentally (e.g. using 23, 24, and 25 hour cycles)

tim *timeless*, a *Drosophila* gene that is required for circadian rhythmicity

time and motion study a systematic method analysis of the operations required to produced manufactured item in a factory; unnecessary motions are eliminated, and more efficient production is achieved

transducer mechanism for converting one type of signal into another type of signal; the pineal gland is a neuroendocrine transducer converting a neural signal to an endocrine signal

transients temporary cycles, usually seen during resetting

twilight the periods between sunrise and sunset and dark; before sunset, civil twilight
    begins the earliest, then nautical twilight, and finally astronomical twilight

Zeitgeber synchronizer, entraining agent, time giver, time signal, time cue.

Societies

International Society for Chronobiology, Dora K. Hayes, Ph.D., Secretary-Treasurer, 9105 Shasta Court, Fairfax, VA 22031, USA

Society for Light Treatment and Biological Rhythms, 10200 W. 44th Ave. #304, Wheat Ridge, CO 30033-2840, telephone 303-424-3697, email: sltbr@recourcenter.com

Society for Research on Biological Rhythms, Gilmer Hall, University of Virginia, Charlottesville, Virginia 22903, clock@virginia.edu.

Journals

Journal of Interdisciplinary Cycle Research (beginning 1970), published by Swets & Zeitlinger (Swets Publishing Service, P.O. Box 825, 2160 SZ Lisse, the Netherlands). Telephone 02521-35111, Fax 02521-15888

Chronobiology International: A Journal of Basic and Applied Biological Rhythm Research (began in 1973), Raven Press. A journal of The International Society of Chronobiology. Michael H. Smolensky, University of Texas, Health Science Center at Houston, 1200 Hermann Pressler Drive, Houston, Texas 77030, Telephone 713-704-58890, FAX 713-704-0583

Journal of Biological Rhythms (began in 1986) Guildford Publications, Inc., 72 Spring Street, New York, NY 10012 (Telephone 800-365-7006).

Software

Wake-up and to-sleep data can be analyzed with **Circadia** software (Thomas A. Houpt, Behavioral Cybernetics, One Kendall Sq. #304, P.O. Box 9171, Cambridge, MA 02139) using the Open Schedule file (MacIntosh). I used Circadia to analyze the wrist activity recordings.

Worldclock is software that shows a world map with dark and light superimposed on it. You can find times in other locations, sunrise and sunset, twilight, and it beautifully illustrates the seasonal effects of light on the earth. Paul Software Engineering, 513 West A Street, Tehachapi, CA 93561-1933, USA.

Action software for analyzing wrist activity is sold by Ambulatory Monitoring, Inc. 731 Saw Mill River rd. Ardsley, NY 10502. 914-693-9240. This company made the wrist monitors used in recording circadian rhythms of freeliving humans shown in this book.

Tau software displays circadian rhythm data and is available from Mini-Mitter Co., Inc. P.O. Box 3386, Sunriver, Oregon 97707, 503-593-8639. Their motto is "Physiological Monitoring for Mice to Men."

Clocks and light sources are sold by **Bio-Brite, Inc**. 7315 Wisconsin Avenue, 900E, Bethesda, Maryland 20814.

ANNOTATED BIBLIOGRAPHY

Annotated Bibliography

ARKING, R. *The Biology of Aging.* Englewood Cliffs, NJ: Prentice-Hall, Inc. Simon & Schuster, 1991. The book covers measurement, evolution, human aging, genetic determinants, and theories of aging. 420 pages.

ASCHOFF, J., ed. *Circadian Clocks.* Amsterdam: North-Holland Pub. Co., 1965. The book has 44 articles from the Feldafing Summer School 1964 sponsored by NATO, including classics about pendulum and relaxation oscillators, and ways to measure phase shifting, and a Glossary in English and German. 479 pages.

ASCHOFF, J., ed. *Biological Rhythms.* vol. 4 of Handbook of Behavioral Biology. 27 invited chapters and Glossary. New York: Plenum Press, 1981. Sections of the book include a survey of rhythms, methods of analysis, mathematical models, general properties of circadian rhythms, freeruns, entrainment, behavioral rhythms, neural and endocrine controls, genetics, development, orientation, tidal and lunar rhythms, circannual rhythms, photoperiodism, short term rhythms, sleep, and ovarian cycles. 562 pages.

ASCHOFF, J., S. DAAN, and G. A. GROOS. *Vertebrate Circadian Systems: Structure and Physiology.* New York: Springer-Verlag, 1982. There are contributions from a meeting in Schloss Ringberg in 1980. 37 articles, 154 figures. The book explores pathways by which Zeitgeber signals reach pacemakers, the role of the SCN and the pineal gland, PRCs and splitting, sleep–wake cycle in humans, and photoperiodism in hamsters and birds. 363 pages.

AYENSU, E. S., and P. WHITFIELD. *The Rhythms of Life.* New York, Crown Publishers, 1981. The coffeetable sized book has pleasing color and black and white photographs and was written for the general public. Chapters cover rhythms of cosmos, season, sex, population, growth, energy, motion, health and disease, motion, and music. There's information and diagrams on how clocks work. 199 pages.

BENNETT, M. F. *Living Clocks in the Animal World.* Springfield, IL: Charles C. Thomas, Publisher, 1974. The book is a readable monograph reviewing biological clocks which focuses on crabs, honey bees, earthworms, mollusks, amphibians. 221 pages.

BINKLEY, S. *The Pineal, Endocrine and Nonendocrine Function.* Englewood Cliffs, NJ, Prentice-Hall, 1988. Anatomy, biochemistry, regulation by light and dark, neural control, mammalian photoperiodism, circadian rhythms, melatonin functions, hormone secretion and action, development, aging, and comparative endocrinology are covered. I have only found one error in this remarkable book. It is illustrated with graphs and photographs. 304 pages.

BINKLEY, S. *The Clockwork Sparrow. Time, Clocks, and Calendars in Biological Organisms.* Englewood Cliffs, NJ: Prentice-Hall, 1990. The book covers freeruns, entrainment, resetting, effects of constant light, photoperiodism, circarhythms, and clock models. A particularly nice chapter reviews the circadian rhythms of commonly studied organisms on a species by species basis. The emphasis in the book is on animal results and it is illustrated with charming photographs, tables, graphs, and in particular, many circadian rhythm records, especially those of house sparrows. This book has the prettiest cover I have ever seen with a wonderful photograph taken by the author. However, the typesetters must have been using poorly focused extraretinal photoreception because it is full of printers errors. 262 pages.

BINKLEY, S. *Endocrinology.* New York: Harper Collins College Publishers, 1995. The college textbook covers the endocrine glands—hypothalamus, pituitary, pineal, retina, thyroid, parathyroid, pancreas, gastrointestinal tract, adrenal, testes, ovary, placenta—and the hormones and related general topics such as aging, hibernation, sleep, cytokines, chalones, erythropoietin, phytohormones, invertebrate hormones, and pheromones. Data showing cycles are presented for most of the hormones. It is printed on the worst paper I've ever seen, but will probably recycle. 539 pages.

Biological Clocks. Vol. XXV of the Cold Spring Harbor Symposia on Quantitative Biology. Baltimore, MD: Waverly Press, 1960. 51 classic symposium papers include the classic, "Circadian Organization," by Colin Pittendrigh with terminology (p. 160) and articles about models, thermoperiodism, photoperiodism, celestial navigation, seasonal migration, lunar, tidal, and annual rhythms. 524 pages.

BLACKMAN, R. B., and J. W. TUKEY. *The Measurement of Power Spectra.* New York: Dover Publications, 1958. The book is written from the point of view of communications engineering. In the sixties, I programmed Fortran representations of the formulae in the book for calculation of time series (autocorrelation, power spectra) on house sparrow circadian rhythm data, random numbers, sine, and square waves. 190 pages.

BRADY, J. *Biological Clocks.* London: Edward Arnold Publishers Ltd., 1979. Seven chapters cover daily, tidal, and annual rhythms, the exogenous or endogenous issue, circadian rhythms, celestial navigation, continuously consulted clocks, photoperiodism, and clock mechanisms. 60 pages.

BRADY, J., ed. *Biological Timekeeping*. New York: Cambridge University Press, 1982. Intended as a student text by authors from a seminar organized by the Society for Experimental Biology at Imperial College in London in 1980. A glossary and eleven invited chapters. 197 pages.

BROWN, F., J. W. HASTINGS, and J. D. PALMER. *The Biological Clock, Two Views*. New York: Academic Press, 1970. A concise discussion of contrasting exogenous and endogenous hypotheses for explaining rhythmic phenomena by their respective proponents. 94 pages.

BÜNNING, E. *Die Physiologische Urh*. Berlin: Springer-Verlag, 1958. The first monograph in the field of circadian rhythms. Its focus is on plants.

BÜNNING, E. *The Physiological Clock* (3rd ed.) New York: Springer-Verlag, 1973. The book is an enduring monograph in English originally published in 1958 under the title *Die Physiologishe Uhr*. The author discusses endodiurnal oscillations, periodicity fate-out, initiation by external factors, controlling organs, temperature, light, relaxation oscillators, biochemistry, synchronization, use of the clock in direction finding, circadian and tidal and lunar rhythms, day-length measurement, splitting in the shrew, beats, damage by constant light, and the absence of synchronizing stimuli. 167 pages.

BURNS, J. R. *Cycles in Humans and Nature, An Annotated Bibliography*. The Scarecrow Press Inc., Metuchen NJ, © 1994. The useful compilation includes sections on astrophysics, atmospheric sciences, biology, botany, economics, geoscience, medicine, social science, and zoology. The book is much more interesting than the phrase "annotated bibliography" suggests. It has an author index. 288 pages.

CARPENTER, D. O., ed. *Cellular Pacemakers*. New York: John Wiley & Sons, Inc., 1982. Vol. 2 of *Function in Normal and Disease States*. 15 invited articles cover cellular pacemakers in the nervous system, circadian rhythms, and what happens when pacemaking systems break down in disease and aging. 371 pages.

COLEMAN, R. M. *Wide Awake at 3:00 A.M. By Choice or by Chance*. W. H. Freeman and Company, New York, 1986. Emphasis on human and sleep related aspects of circadian rhythms, Glossary. 195 pages.

CONROY, R., and J. MILLS. *Human Circadian Rhythms.* London: J. & A. Churchill, 1970. Methods, abnormal time schedules, and rhythms (endocrine, temperature, kidney, cardiovascular, respiratory, wakefulness, birth). The book opens with a section discussing definitions and has 945 references. 236 pages.

DECOURSEY, P. J., ed. *Biological Rhythms in the Marine Environment.* Columbia, SC: University of South Carolina Press, 1976. Twenty research articles from a symposium at the Hobcaw House on the Belle W. Baruch Coastal Research Station near Georgetown. Articles on decapod crustaceans, gastropods, mice, crabs, marine algae, flatfish, and amphipods include vertical migration rhythms and celestial orientation and methods of rhythm analyses. 283 pages.

DINGES, D. F., and R. J. BROUGHTON. *Sleep and Alertness: Chronobiological, Behavioral, and Medical Aspects of Napping.* New York: Raven Press, 1989. 27 contributed papers characterize napping, napping and sleep, shift work, siestas, and sleepiness.

EDMUNDS, L. N. *Cell Cycle Clocks.* New York: Marcel Dekker, Inc., 1984. 27 contributed articles detail cell division cycles and their temporal organization and including the cancer pathology, cancer chronotherapy, and aging. 616 pages.

EDMUNDS, L. N. *Cellular and Molecular Bases of Biological clocks: Models and Mechanisms for Circadian Timekeeping.* New York: Springer-Verlag, 1987. The book covers eukaryotic microorganisms, cell cycle clocks, experimental approaches, biochemical models, molecular models. There are 156 illustrations, a list of abbreviations, 57 pages of references, an author index and a subject index. 497 pages.

EHRET, C. F., and L. W. SCANLON. *Overcoming Jet Lag.* New York: Berkley Pub. Corp., 1983. A pocket sized book gives the details for diet and light prescriptions in a 3-step program (preflight, inflight, and post-flight steps) using light, feasting, fasting, and caffeine. The program was developed at the Argonne National Laboratory for executive travelers and the U.S. Army rapid deployment forces hoping to reduce the time shifting effects of east–west travel. 160 pages.

ENRIGHT, J. T. *The Timing of Sleep and Wakefulness: On the Substructure and Dynamics of the Circadian Pacemakers Underlying the Wake–Sleep cycle.* A monograph with computer simulations. New York: Springer-Verlag, Inc., 1980. The monograph includes computer simulations, envisions an ensemble of coupled relaxation oscillators, and has a particularly useful and unique discussion of precision of biological rhythms. 263 pages.

FOLLETT, B. K., and D. E. FOLLETT. *Biological Clocks in Seasonal Reproductive Cycles*. New York: John Wiley & Sons, 1981. The book has 20 papers, is the Proceedings of the 32nd Symposium of the Colston Research Society held in the University of Bristol in 1980, emphasizes photoperiodism, and has both a species and a general index. 292 pages.

GLASS, L. *From Clocks to Chaos: The Rhythms of Life*. Princeton: Princeton University Press, 1988.

GRIFFIN, D. R. *Bird Migration*. New York: Dover Publications, 1974. The book explains bird migration and how it is studied using vanishing points and bands and radar and airplanes, the seasonal timing of migrations, how the energy requirements are met with fattening, how birds navigate (thermal radiations? terrestrial magnetism? earth's rotation?), pigeon homing, migratory restlessness. 180 pages.

GWINNER, E. *Circannual Rhythms: Endogenous Annual Clocks in the Organization of Seasonal Processes*. New York: Springer-Verlag, 1980. The book is an inventory of work on the subject of circannual rhythms, their freeruns, their synchronization, their interrelationships, and their adaptive significance. There is a systematic index, a useful table of investigations of circannual rhythms organized by species, and 73 figures of scientific data and models. The book is volume 18 in the series of Zoophysiology. 263 pages.

HALBERG, F., ed. *Proceedings of the XII International Conference of the International Society for Chronobiology*. Milano, Italia: Il Ponte, 1975. Nearly 100 research articles from a 1975 conference held in Washington, D.C. 782 pages.

HALBERG, F., F. CARANDENTE, G. CORNELISSEN, and G. S. KATINAS. *Glossary of Chronobiology*. Milano, Italia: Il Ponte, 1977. English/Italian. One or more pages is devoted to the explication of the terminology including acrophase, autorhythmometry, frequency, chronobiology, circaseptan, least-squares method, and many others. 188 pages.

HARKER, J. *The Physiology of Diurnal Rhythms*. New York: Cambridge University Press, 1964. The book is a monograph covering the role of the environment, freerunnning rhythms, phase shifting, and rhythm abnormalities. 114 pages.

HIROSHIGE, T., and K. HONMA, eds. *Circadian Clocks and Zeitgebers*. Sapporo, Japan: Hokkaido University Press, 1985. There are 17 research articles, Proceedings of the first Sapporo Symposium on Biological Rhythm, August 29–31, 1984. They cover *in vivo* and *in vitro* circadian clocks of bread mold and quail, suprachiasmatic nuclei transplants, entrainment in crickets and rats, and the relationship of the maternal rhythms to those of their pups in rats. 191 pages.

KALES, A., ed. *Sleep, Physiology & Pathology*. Philadelphia: J. B. Lippincott Co., 1969. The book resulted from a symposium in Los Angeles in 1968 and covers the physiology, pathology, and drugs of sleep. 360 pages.

KARMANOVA, I. G. *Evolution of Sleep*, New York: Karger, 1982. The book was translated from Russian. It discusses sleep in fish, amphibians, reptiles, birds, and mammals, and the evolution of the circadian biorhythmicity of sleep from a primary sleep to a more complex form with paradoxical sleep. 164 pages.

KRIEGER, D. T. *Endocrine Rhythms*. New York: Raven Press, 1979. 12 invited chapters consider the classification, causation, and properties of endocrine rhythms and the neuroanatomy of the pathways involved in their regulation. 332 pages.

KUPFER, D. J., T. H. MONK, and J. D. BARCHAS. *Biological Rhythms and Mental Disorders*. New York: Guilford Publications, Inc., 1988. The book has 10 invited chapters dealing with depression, sleep disturbances, circadian rhythms, and circannual rhythms. 357 pages.

LAMBERG, L. *Bodyrhythms: Chronobiology and Peak Performance*. New York: William Morrow and Company, 1994. The author of the self-help book is a medical journalist. There is a glossary and a list of shift work consulting firms. There is a foreword by well known sleep researcher, William C. Dement, M.D., Ph.D., a Glossary, an appendix of strategies for self-help, an Owl/Lark self-help test, regularization of the menstrual cycle with light, jet lag advice with a useful time zone guide, instructions for a personal sleep/wake diary, a sleepiness/alertness chart, and a list of shift work consulting firms. 274 pages.

LEVERT, S. *Melatonin: The Anti-Aging Hormone*. New York: Avon Books, 1995. This self help book is written in question and answer format. The disclaimer warns that the procedures are not meant to replace medical professional advice and disclaims liability. 226 pages.

LIBERMAN, J. *Light: Medicine of the Future, How We Can Use It to Heal Ourselves NOW*. Santa Fe, NM: Bear & Company, 1991. Dr Liberman, an Aspen optometrist, attempts to marry the intuitive and the rational sciences with light. Topics include the role of colors (pink for prisoners, red for athletes), UV radiation, and sunglasses. There are illustrations, 20 in lovely color. 251 pages.

LOFTS, B. *Animal Photoperiodism*. London: Edward Arnold Pubs. Ltd., 1970. The concise book has nice photographs the sparrow biology. It is No. 25 in the Institute of Biology's *Studies in Biology* series. 62 pages.

LUCE, G. G. *Biological Rhythms in Psychiatry and Medicine*. U.S. Government Printing Office, Public Health Service Publication No. 2088, 1970. A National Institute of Mental Health report containing about human temporall organization and its implications for mental and physical health. 183 pages.

LUCE, G. G. *Body Time: Physiological Rhythms and Social Stress*. New York: Bantam Books, Inc., 1971. A book for the general audience derived from the NIMH report. 441 pages.

MENAKER, M., ed. *Biochronometry*. Washington, D.C.: National Academy of Sciences, 1969. Proceedings of a symposium at Friday Harbor, Washington, in 1969. 40 research articles "that analyze circadian systems at the physiological level and that lead one to expect further rapid progress." 662 pages.

MENDLEWICZ, J., and H. M. VAN PRAAG. *Biological Rhythms and Behavior*. New York: S. Karger, 1983. The book is Volume 11 of Advances in Biological Psychiatry. The book has 13 articles, mostly about people (some rats and monkeys) by different authors, 63 figures and 16 tables. 149 pages.

MEYERSBACH, H. VON, L. E. SCHEVING, and J. E. PAULY, eds. *Biological Rhythms in Structure and Function*. New York: Alan R. Liss, 1981. 13 papers from the International Congress of Anatomy held in Mexico City in 1980. 241 pages.

MILLS, J. N. ed. *Biological Aspects of Circadian Rhythms*. London: Plenum Press, 1973. Glossary and eight reviews by nine authors. It includes a discussion of latitude and the humans, circadian rhythms of parasites, and bird migration and orientation. 319 pages.

MOORE-EDE, M. C., and C. CZEISLER. *Mathematical Models of the Circadian Sleep–Wake Cycle*. New York: Raven Press, 1984. Papers by eight authors are collected in this book. 216 pages.

MOORE-EDE, M. C., F. M. SULZMAN, and C. A. FULLER. *The Clocks That Time Us.* Cambridge, MA, Harvard University Press, 1982. Introductory book with Glossary. 448 pages.

MOORE-EDE, M., S. CAMPBELL, and R. J. REITER. *Electromagnetic Fields and Circadian Rhythmicity.* Boston: Birkhauser, 1992. The book, contributed papers from a conference supported by a contract from the Electric Power Research Institute (EPRI), discusses the effects of extremely low frequency electromagnetic radiation on organisms, particularly upon their circadian rhythms. 210 pages.

NELSON, R. J. *An Introduction to Behavioral Endocrinology.* Sunderland, MA: Sinauer Associates, Inc., Pubs. 1995. The book has a chapter on Biological Rhythms and Behavior, pp. 381–442. 611 pages.

OREN, D. A., W, REICH, N. ROSENTHAL, and T. WEHR. *How to Beat Jet Lag, A Practical Guide for Air Travelers.* New York: Henry Holt and Company, 1993. The self help book is organized by giving recommendations for light exposure adjustments for flights on journeys longer than three days by number of time zones crossed, east or west, is sold with a cloth eyemask and Klingers® eyeshades. 131 pages.

PALMER, J. D. *Biological Clocks in Marine Organisms: The Control of Physiological and Behavioral Tidal Rhythms.* New York: John Wiley & Sons, Inc., 1982. The book covers oscillations with a period of about 24.8 hours, vertical migration rhythms, color-change rhythms in fiddler and green crabs, experiments where tidal rhythms are studied after translocation across time zones, and has a Glossary. 173 pages.

PALMER, J. F., F. BROWN, and L. EDMUNDS. *An Introduction to Biological Rhythms.* New York: Academic Press, 1976. Intended as a textbook for biologists, it has a glossary. 375 pages.

PAULY, J. E., and L. SCHEVING. *Progress in Clinical and Biological Research.* Two Volumes: 227A and 227B. Advances in Chronobiology, Part A., Proceedings of the XVIIth International Conference of the International Society for Chronobiology. New York: Alan R. Liss, 1987. 613 pages.

PAVLIDIS, T. *Biological Oscillators: Their Mathematical Analysis.* New York: Academic Press, Inc., 1973. The monograph is about "biological oscillators and the mathematical techniques used for investigation of biological oscillators. Uses elementary differential equations and linear algebra. 207 pages.

PENGELLEY, E. T. *Circannual Clocks: Annual Biological Rhythms.* New York: Academic Press, Inc., 1974. 14 papers comprising the proceedings of a Satellite Symposium of the 140th Meeting of the American Association for the Advancement of Science in San Francisco. Includes republication of a classic paper by William Rowan showing that daily light increases change the sex organs of Juncos. 523 pages.

PIERPAOLI, W. and W. REGELSON. *The Melatonin Miracle: Nature's Age-Reversing, Disease-Fighting, Sex-Enhancing Hormone.* New York: Simon and Schuster, 1995. The book is written for the general audience and reprints the original articles on the mouse research. 255 pages.

PORTER, R., and G. M. COLLINS, eds. *Photoperiodic Regulation of Insect and Molluscan Hormones.* London: Pitman, 1984. The book linking "wet" and "dry" physiology has 18 research articles on invertebrate hormones. Articles deal with flesh-flies, fruit flies, cockroaches, moths, blood feeding bugs, aphids, spider mites, and slugs. The articles derive from the symposium on Photoperiodic Regulation held at the Ciba Foundation in London in 1983. 298 pages.

REINBERG, A., and M. H. SMOLENSKY. *Biological Rhythms and Medicine: Cellular, Metabolic, Physiopathological, and Pharmacologic Aspects.* New York: Springer-Verlag, 1957. There are seven chapters by seven authors, intended as a textbook of modern applied chronobiology, with 148 illustrations. 308 pages.

RICHTER, C. P. *Biological Clocks in Medicine and Psychiatry.* Springfield, IL: Charles C. Thomas, Publisher, 1965. The book, a classic, originated from the author's lectures on biological clocks in animals and in humans. Richter is credited for his skillful recording of free-running rhythms, exclusion of the endocrine glands as the sites of the clock, and the first work showing that a small area of the hypothalamus was a pacemaker. The small area was subsequently identified as the suprachiasmatic nuclei. 109 pages.

ROSE, K. J. *The Body in Time.* New York: John Wiley & Sons, Inc., 1988. A science writer considers a potpourri of time courses for biological events including but not limited to sleep and circadian rhythms (sneezes, rates of development, development of an embryo). He divides temporal physiology into chapters milliseconds, seconds, minutes, hours, days, months, and years. 237 pages.

ROSENTHAL, N. E., and M. C. BLEHAR, eds. *Seasonal Affective Disorders & Phototherapy.* The Guilford Press, 1989. The book is a collection of 20 papers by researchers and clinicians. It includes work on humans and the search for animal models, seasonal changes in the normal population, and depression. 386 pages.

ROSENTHAL, N. E., M.D. *Winter Blues: Season Affective Disorder, What It Is and How To Overcome It*. New York: The Guilford Press, 1993. The author's own experiences add interest to this self help book which explains seasonal depression and light therapy. There's a list of where to get further personal help and a Self-Assessment Mood Scale for SAD—even menus and recipes for oat bran muffins, very spicy chick peas, lentil spaghetti sauce, sweet and sour apricot chicken or fish, and curried barley and lentil soup. 326 pages.

ROSENTHAL, N. E., M.D. *Seasons of the Mind: Why you get the winter blues & what you can do about it*. The Guilford Press: New York, 1989. Literary quotes lighten this book about SAD. There is a list of places that specialize in SAD and diet and recipe advice. Dr. Rosenthal was chief of the Section on Environmental Psychiatry, Clinical Psychobiology Branch, Intramural Research Program, at the National Institute of Mental Health (NIMH), and director of light therapy studies at NIMH when he wrote the book. 278 pages.

SAUNDERS, D. S. *A Concise Introduction to Biological Rhythms*. Glasgow and London: Blackie, 1970. The book is aimed for advanced, presumable Scottish, undergraduates and those beginning a research career in biochronometry and has a glossary. 170 pages.

SAUNDERS, D. S. *Insect Clocks*. Oxford: Pergamon Press, 1976. The monograph covers circadian rhythms and seasonal photoperiodism in insects, primarily flies. 279 pages.

SCHEVING, L. E., F. HALBERG, and J. E. PAULY, eds. *Chronobiology*. Tokyo: Igaku Shoin LTD., 1974. Proceedings of the first conference of the International Society for Chronobiology held in Little Rock, Arkansas. 782 pages.

SMYTH, A. *Seasonal Affective Disorder—who gets it? what causes it? how to treat it?* London: Thorsons, 1991. Ms. Smyth is a medical journalist. There is a list of SAD research centers and lighting sources in this self-help book. 275 pages.

SOLLBERGER, A. *Biological Rhythm Research*. Amsterdam: Elsevier Pub. Co., 1965. 461 pages.

STETSON, M. H., ed. *Processing of Environmental Information in Vertebrates*. New York: Springer-Verlag., 1988. The book of eleven chapters arose from a symposium of the Division of Comparative Endocrinology of the American Society of Zoologists. Papers discuss fish, lizards, amphibians, birds, hamsters, the mammalian fetus, photoperiodism, and reproductive function. 261 pages.

STROGATZ, S. *The Mathematical Structure of the Human Sleep–Wake Cycle*. New York: Springer-Verlag, 1986. Using human raw data obtained from other investigators and presented as a data bank of raster plots, the author analyzes some models and makes some simulations which qualitatively fit the data, but which, he says, fail more rigorous tests. 230 pages.

STRUGHOLD, H. *Your Body Clock*. New York: Charles Scribner's Sons, 1971. The book is about physiological rhythm responses to rapid environmental changes and discusses sleep, shift work, air travel, and space missions. 94 pages.

SWEENEY, B. M. *Rhythmic Phenomena in Plants*. New York: Academic Press, 1969. The book captures the excitement of the experiments written for a student who has heard about rhythms and asked what they are and how they work. 147 pages.

VIDRIO, E. A., ed. *Progress in Clinical and Biological Research Volume 59C: Biological Rhythms in Structure and Function*. New York, Alan R. Liss, Inc., 1981. Thirteen papers from a symposium of the Eleventh International Congress of Anatomy, held in Mexico City in 1980. Articles cover rhythms in tissues and cells, chronohistochemistry, rhythms in susceptibility, circannual rhythms, and control of rhythms in humans, mice, hamsters, and rats including interesting rhythms in the appearance of rat liver cells. 241 pages.

WALKER, C. A., C. M. WINGET, and K. F. A. SOLIMAN. *Chronopharmacology and Chronotherapeutics*. Tallahassee, FL: Florida A & M foundation, 1981. 39 articles are the proceedings of the International Symposium on Chronopharmacology and Chronotherapeutics held in Tallahassee, Florida. Drug administration techniques, neuropharmacology, cardiovascular drugs, chemotherapy, endocrinology, immunology, toxicology, parasitology, and antiemetics are discussed. 417 pages.

WARD, R. R. *The Living Clocks*. New York, Mentor, 1971. The book is aimed at readers with no science background.

WEHR, T. A., and F. K. GOODWIN, eds. *Circadian Rhythms in Psychiatry, V. 2 of Psychobiology and Psychopathology*. Pacific Grove, CA: The Boxwood

Press, 270 pages. Thirteen research articles about humans having to do with affective disorders, depression, mania, light, antidepressant drugs, and the pineal gland. 351 pages.

WAUGH, A. E. *Sundials, Their Theory and Construction.* New York, Dover Publications, Inc., 1973. This fascinating little book explains kinds of time, noon marks, equatorial sundials, horizontal sundials, north and south and east and west sundials, polar dials, vertical delining dials, direct reclining dials, analemmas, ceiling dials, dial furniture, portable dials, armillary spheres, and memorial dials. He explains how you can tell time by stepping off the length of your own shadow, and odd trick, and not so easy to do. 228 pages.

WEVER, R. A. *The Circadian System of Man: Results of Experiments Under Temporal Isolation.* Springer-Verlag, New York Inc., 1979. This book of scientific research contains 181 illustrations compiled from experiments done over a dozen years at the Max-Planck-Institut für Verhaltensphysiologie in Andechs including many which demonstrate the "internal clock" of individual human beings. Topics include dissociation, desynchronization, light intensity, temperature, electromagnetic fields, and oscillator models. 276 pages.

WINFREE, A. T. *The Geometry of Biological Time.* New York: Springer-Verlag, 1980. The book is Volume 8 of Biomathematics series and has 290 illustrations. 530 pages.

WINFREE, A. T. *The Timing of Biological Clocks.* Scientific American Library, 1987. The book has 233 illustrations including color computer graphics. 199 pages.

WURTMAN, R. J., M. J. BAUM, and J. T. POTTS, JR. *The Medical and Biological Effects of Light.* New York: New York Academy of Sciences, 1985. The book is volume 43 of the Annals of the New York Academy of Sciences, a compilation of papers from a conference held in New York City in 1984. Photobiology, photosensitization. photoperiodism, biological rhythms, and impact of interior lighting on health are covered. 408 pages.

SECTION III: MEASURE YOUR OWN CIRCADIAN RHYTHM

Measure Your Own Circadian Rhythm

1   Getting Started, Needful Things

How is this section organized?

This section consists of a series of charts and forms with blanks for information. Instructions are provided for filling out the charts and forms. Examples of data from other individuals is provided for you to compare yourself with others. Getting started is easy.

Needful things:

✓   You need a **pencil**.
✓   You need a blank copy of the **wake-up to-sleep form**.
✓   You need your **clock** or wristwatch.

I found it easiest to use a bedside clock and to keep my chart on my night table. I wrote down the time I woke up. I wrote down the time I turned out the lights to go to sleep, which for me, was usually only a few minutes before nodding off. After a week, where I sometimes forgot, it was a habit.

●   Don't give up!
●   Even a few days of your own data will be interesting.
●   If you start to keep records, then forget or stop for some reason, don't throw away your data. Just start again from whatever date you are on.
●   A couple of days of record is worthwhile, a week is better, four weeks is best.
●   You may think this is a great deal of trouble, and it is, and that someone should invent someone to do it automatically, someone has. (See Figures)

2   Wake-up To-sleep Form, Instructions

Use a clock set to local time. Write the times you wake-up and the times you go to-sleep in the spaces on the form.

- **date**, month/day/year. Normally you should use one line for each date. If however, you have a broken up sleeping schedule or are doing shiftwork so that several wake-ups or to-sleeps occur in the same 24 hour period, you may want to use more lines for a date
- **day**, days of the week (Sunday, Monday, Tuesday, Wednesday, Thursday, Friday, Saturday)
- **wake-up**, the time a.m. or p.m. you woke up and stayed up (e.g. 7:55 a.m.)
- **to-sleep**, the time (a.m. or p.m.) you go to sleep, and stayed asleep (e.g. 12:30 a.m., half past midnight)
- **notes**, space is provided for noting times of events (e.g. an alarm woke you up, a journey, you changed time zones by *x* hours when traveling, you were menstruating, you were sick, you stayed up late to study for an exam or to enjoy a social event, the baby woke you up, you were called in to work at night, you took a nap, you changed work shifts, etc.).

Do you have more than one wake-up and/or to-sleep time each day? If you woke up or went to sleep more than one time on a given day (e.g. because of naps or shift work) you can use a separate line for each wake-up and to-sleep.

Put the to-sleep on the same line as the preceding wake-up, even if the to-sleep is on the next day of the week.

If you are keeping your record while traveling across time zones, your record will be more difficult to keep track of, but it will be more interesting. You should look through this section for an alternate method of keeping track of your cycles.

Accuracy to the nearest five minutes was the norm I tried to achieve, but accuracy to the nearest fifteen minutes should be good enough.

Wake-up To-sleep Form, Copy 1.

Date	Day of Week	Wake-up	To-sleep	Notes
5/19/90	SAT	7:55 a.m.	1:30 a.m.	Example

Wake-up To-sleep Form, Copy 2.

Date	Day of Week	Wake-up	To-sleep	Notes
5/19/90	SAT	7:55 a.m.	1:30 a.m.	Example

3   Conversion to Decimal Equivalent Time

I defined DEQ as the decimal equivalent of Eastern Daylight Savings time.

Why should you convert your data to decimal equivalents? It is difficult to make averages with data in minutes, but it is simple to do calculations if the data are in decimal form. Enough space is provided on the DEQ form for a month of data, but you need not necessarily begin on the first of the month. Day 1 should be the first day you started recording (**date** = month, day, year, **day** = Monday, Tuesday, Wednesday, Thursday, Friday, Saturday, Sunday, **wake-up** = wake-up in DEQ, **to-sleep** = to-sleep in DEQ, **awake** = to-sleep minus wake-up).

You need to choose a home time zone. For data analysis you will eventually have to convert all your times to a decimal equivalent of time in one time zone. To do this you will ultimately have to choose a "home time zone". Unless you are traveling, a convenient choice is the local time you are observing right now. The subjects in the examples provided in "How Do You Measure Up?" used Eastern Daylight Savings Time as their home time zone was in the Delaware Valley of Pennsylvania.

● To convert the hours, determine the number of hours since the midnight before the last wake-up. If you retired at 1 a.m., there will be 25 hours since the previous midnight. Otherwise the number of hours will be between 0 and 24.

● To convert the minutes to decimal form, divide the minutes you recorded by 60 (e.g. 30 minutes divided by 60 = 0.50).

The reason you are doing this becomes apparent when you try to make the graph, your Circadian Rhythm Time Chart.

Sample Conversions to DEQ.

Wake-up	To-sleep
4:00 a.m. = 4.0 DEQ	9:00 p.m. = 21.0 DEQ
8:13 a.m. = 8.2 DEQ	11:18 p.m. = 23.3 DEQ
noon = 12.0 DEQ	Midnight = 24.0 DEQ
1:48 p.m. = 13.8 DEQ	1:30 a.m. = 25.5 DEQ

DEQ Form, Copy 1.

Date	Day of Week	Wake-up	To-sleep	Notes
5/19/90	SAT	7:9	25.5	Example

DEQ Form, Copy 2.

Date	Day of Week	Wake-up	To-sleep	Notes
5/19/90	SAT	7:9	25.5	Example

4   Circadian Rhythm Time Chart

Your circadian rhythm time chart will show you the status of your natural daily rhythm. This rhythm is innate to your individual body. Normally, this rhythm is synchronized to 24 hours with your environment—by light, social cues, alarm clocks, and wristwatches. When you change time zones, do shiftwork, change between stand-ard and daylight savings time, or undergo some other change in your sleep–wake schedule, these changes will show up in the pattern you display. Your pattern will represent the rhythm of time awake. The Circadian Rhythm Time Chart will display your pattern.

●   To fill out the chart, using a pencil, blacken the space between the time you wake up and the time you go to sleep—your awake time—for each date.

●   Use the data in the DEQ form.

Circadian Rhythm Time Chart, Copy 1.

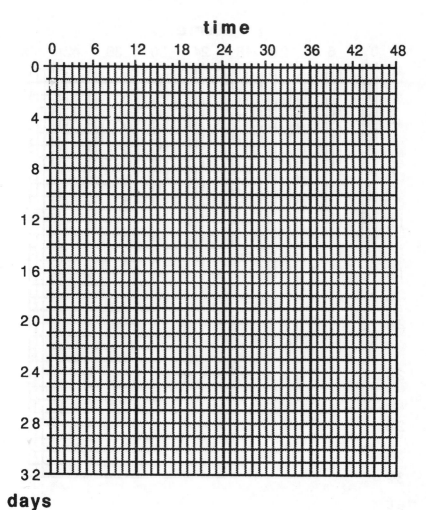

Circadian Rhythm Time Chart, Copy 2.

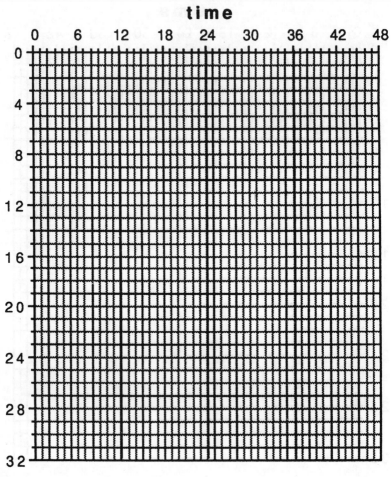

time

days

5   Day of the Week

● Use the data you have converted to DEQ (DEQ form).

● Exclude naps.

● Make an average for all the Monday wake-ups and record it on the day of the week form; make an average for all the Tuesdays, etc.

● Repeat the procedure for all the to-sleeps.

● To calculate the awake column, subtract the wake-up value from the to-sleep value.

● If you collected data for a month (one complete DEQ form), you should have about 4 values to average for each space in the day of the week form. However, if you have less data, you can still arrange it by day of the week.

● If you did shiftwork, you may have more than one wake-up or to-sleep on each day which means you cannot calculate valid averages.

● For naps, you can calculate your nap data separately from your main sleeping period. For shiftwork, you can calculate the data separately for the days spent on each shift.

Day of the Week Form.

	Wake-up	To-sleep	Awake
Monday			
Tuesday			
Wednesday			
Thursday			
Friday			
Saturday			
Sunday			

6 Grand Means and Ranges

If you have recorded wake-up and to-sleep times (even for a few days), did not travel across time zones, and did not do shift work, you can make some averages (means) and fill out the grand means and ranges form.

To calculate your grand means, determine the average wake-up and to-sleep times using data on the wake-up to-sleep DEQ form.

● Calculate awake (to-sleep DEQ minus wake-up DEQ) and asleep (24.0 minus awake).

● Convert the DEQ means to local time (home time zone).

● Look through the wake-up and to-sleep DEQ data on your DEQ form and find the earliest wake-up and to-sleep times (lowest values). Record them under minimum wake-up DEQ and to-sleep DEQ.

● Look through the awake column and find the shortest awake time and record it. You can *estimate* the shortest asleep times by subtracting the awake maximum from 24. You can convert the wake-up and to-sleep values to local time.

● Look through the wake-up and to-sleep DEQ data on your DEQ form and find the latest wake-up and to-sleep times (highest values). Record them under maximum wake-up DEQ and to-sleep DEQ.

● Look through the awake column and find the longest awake times and record it.

You can *estimate* the longest asleep times by subtracting the minimum awake time from 24. (To calculate actual asleep times, you would have to determine the hours from to-sleep to the next day's wake-up.) You can convert the wake-up and to-sleep values to local time.

Grand Mean and Ranges Form.

	Mean	Minimum	Maximum
Wake-up DEQ hours			
To-sleep DEQ hours			
Awake hours			
Asleep * (24 − Awake) hours			
Wake-up Local Time **			
To-sleep Local Time **			

* For subjects with awake times in excess of 24 hours, asleep times may not make sense calculated this way.
** You may wish to convert your times back to local time.

7 Interpreting Your Results

Look at your Circadian Time Chart. This method of plotting your daily cycles shows your daily pattern of activity and rest. You can see the distribution of your activity with reference to time of day (horizontal axis, time of day). You can see the changes in your activity pattern on a day-to-day basis over the course of time (vertical axis, e.g. a month).

The figures show the author's own results: weekend delay, particularly in the author's wake-up time; that the author was "entrained;" and the effect of changing between daylight savings and standard time.

A table shows the wake-up and to-sleep data for eleven people. The subjects remained in their home time zone (Eastern Daylight Savings Time, September–October) and collected data for a month. They did not travel, they did not do shiftwork, they did not change between daylight savings and standard time. The numbers at the bottom are the grand means. The earliest riser in the group of eleven people woke up at 6.77 DEQ (6:46 a.m. Eastern Daylight Savings Time) and the individual who went to sleep the latest retired at 24.65 DEQ (39 minutes after midnight, Eastern Daylight Savings Time). The average wake-up time was 7.62 DEQ (7:39 a.m.) and the average to-sleep time was 23.95 DEQ (11:57 p.m.). The individual with the longest awake time was awake 1.67 hours more than the individual with the shortest awake time.

A group of three older subjects (over 44 years old) was awake longer, 17.63 hours.

To evaluate your own day of the week data, look at your "day of the week" form. Which day did you wake up earliest? Which night did you go to sleep latest? Which day were you awake longest? The 11 people woke earliest on Wednesday, went to sleep latest on Friday and Saturday, and were awake longest on Friday. They had a five day work week (Monday–Friday) and weekends (Saturday, Sunday). You probably also have a "weekly" cycle.

Wake-up and To-sleep Data for 11 People.

	Wake-up	To-sleep	Awake
1	8.13	24.55	16.42
2	6.92	24.42	17.50
3	8.15	24.24	16.12
4	6.77	23.63	16.86
5	8.42	24.26	15.84
6	7.02	23.05	16.03
7	7.69	23.60	15.91
8	7.81	24.13	16.32
9	8.82	24.65	15.83
10	7.09	23.57	16.48
11	7.01	23.27	16.26
Mean	7.62	23.95	16.32

Day of the Week Data for 11 People.

Day	Wake-up Time DEQ	To-sleep Time DEQ	Awake Hours
Monday	6.9	23.5	16.6
Tuesday	7.6	23.5	15.9
Wednesday	6.6	23.7	17.1
Thursday	7.6	23.9	16.3
Friday	6.9	24.6	17.7
Saturday	8.9	24.6	15.7
Sunday	9.2	23.7	14.5

8 Travel

Keeping track of your rhythm during travel across time zones is a little more demanding.

I recommend recording your rhythm every day for one or two weeks before travel, during the travel and at the destination, and then for one or two weeks following the travel.

A different Wake-up To-sleep Form has been provided for travel with columns added so that you can note the zone for which the wake-up and to-sleep are being recorded.

When you get home, before making a Circadian Rhythm Chart, convert all the times to the decimal equivalent of your home time. Add or subtract the hours of the time change to the data for the destination. In other words, all the data plotted on the Circadian Rhythm Chart should be in the times for one zone. This permits you to view the action of your biological clock with respect to time and, if you traveled across time zones, you should see some phase shifts and possibly some transients as you adapt to times in new zones.

The form is also useful if you want to measure your rhythm during a change between daylight and standard time. If you are using daylight time for the home time zone, then 6:00 p.m. home time is equal to 5:00 p.m. standard time.

Wake-up To-sleep Form (More Than One Time Zone).

Date	Day	Wake-up	Wake zone	To-sleep	Sleep zone	Notes
5/19/90	SAT	7:55 a.m.	DST	1:30 a.m.	DST	Example

9   Shift Work

If you are doing shift work, I would recommend trying to keep a long record recording your rhythm every day as you work through all your various shifts. Use the notes column to write down what shift you were on for each day that you record data.

10   Analyze Your Data

Raster graphs of your data can also be made with Microsoft Excel or with Cricketgraph (Hint: With Cricketgraph, plot lines vertically for each day, turn the graph sideways to see the raster).

You probably have other graphic software and with a little fiddling can make a raster graph of your rhythm. The trick is usually to plot each waking period as a separate line defined by two points (wake-up and to-sleep time).

ABOUT THE AUTHOR

About the Author

The author investigated circadian rhythms (in body temperature, in locomotor activity, in enzymes, in hormones) of humans and animals (sparrows, chickens, hamsters, rats, mice, tadpoles, lizards) and she studied other rhythms (human menstrual rhythms) for over a quarter of a century. She earned her B.A. at the University of Colorado (Boulder, biology), attended Rice University (Houston, physiology), earned her Ph.D. at the University of Texas (Austin, physiology), did postdoctoral work at the National Institutes of Health (Bethesda, physiology), and taught biological clocks and endocrinology for 22 years. Her research was supported by the National Science Foundation, by the National Institutes of Health, and by Temple University. She is author of three textbooks—*The Pineal: Endocrine and Nonendocrine Function* and *The Clockwork Sparrow: Time, Clocks and Calendars in Biological Organisms* (Prentice-Hall, published in 1988 and 1990), and *Endocrinology* (HarperCollins ©1995). She published over 90 scientific articles on circadian rhythm research, and she published some of her experiments in *Scientific American* ("A timekeeping enzyme in the pineal gland." volume 240, pages 50–55, April 1979). The author has also written a column, Mscientia, for America Online (Keyword: Women).

Solebury, PA; 4/7/96 (first day of Eastern Daylight Savings Time)–8/25/96.

NOTES AND REFERENCES

Notes and References

1. Burns, J. *Cycles in Humans and Nature.* (Metchen, N. J., The Scarecrow Press, Inc., 1994).
2. The kind professor was Dr. Michael Menaker of the University of Virginia whose students were then, circa 1968, in Austin, investigating the circadian rhythms and the pineal gland in house sparrows. I had previously done graduate work at Rice University in Houston and met the Western Diamond Back rattlesnake while studying pack rats in southwest Texas with hibernation investigator, Dr. Jack Hudson.
3. Hiroshige, T. *Circadian Clocks and Zeitgebers* (Sapporo: Hokkaido University Press, 1985) pp. i–iii.
4. Aschoff, J., ed. *Biological Rhythms: Vol 4, Handbook of Behavioral Neurobiology* (New York: Plenum Press, 1981), p. vii.
5. MacGregor, E. *Miss Pickerell Goes To Mars* (New York: McGraw-Hill, 1951).
6. Apparently the order of building is disputed; the dates are those of Richard Atkinson c1978, Balfour, M. *Stonehenge and Its Mysteries* (New York: Charles Scribner's Sons, 1979), pp. 106–10.
7. The azimuth is the horizontal angular distance from a reference direction. American Heritage dictionary. p. 131.
8. Hawkins, G. S. and J. B. White. *Stonehenge Decoded* (New York: Dell Publishing, New York, 1965); C. A. Newham. *The Astronomical Significance of Stonehenge* (Gwent, Wales: Moon Publications, 1972); G. S. Hawkins, *Nature*, June 27, 1964; C. Chippindale. *Stonehenge Complete* (London: Thames and Hudson, 1983).
9. My suggestion that a "priestess" might have been involved in the creation of Stonehenge evoked the comment, from an anonymous reviewer, that priestesses were unlikely because of "male dominated society… [that I did not point to a] a recent simple society ruled by such a sophisticated female elite… [and] by contrast such male elites are well known." The reviewer also referred to me as Ms. instead of Dr. Well, at least I wasn't Missed.
10. Ayensu, E. S. and P. Whitfield. *The Rhythms of Life* (New York: Crown Publishers, 1981), p. 176.
11. If you want to see a "Stonehenge" there is a concrete replica of the Stonehenge in England that was built 1918–29 by Col. Samuel Hill near the Columbia River at Maryhill in the state of Washington. America's Stonehenge at Mystery Hill in North Salem, New Hampshire, is not a replica of the British Stonehenge, but does have stones and sighting lines.
12. Waugh, A. E. *Sundials: Their Theory and Construction* (New York: Dover Publications, Inc., 1973).
13. Shakespeare, W. *Julius Caesar*, Act II, Scene ii.
14. Colin Fletcher. *The Man Who Walked Through Time* (New York: Random House Vintage Books, 1967), p. 214.

15. Margaret Atwood, *The Robber Bride* (New York: Bantam Books, New York, ©1993), p. 522. "Time is not a solid, like wood, but a fluid, like water or the wind. It doesn't come neatly cut into even sized lengths, into decades and centuries. Nevertheless, for our purposes we have to pretend that it does. The end of any history is a lie in which we all agree to conspire."
16. Stephen King, *The Green Mile #4, The Bad Death of Delacroix*, (Signet, Penguin Group, New York, 1996), p. 7.
17. Jordan, M. *Encyclopedia of Gods* (New York: Facts On File, Inc., 1993), p. 108.
18. *Random House Compact Unabridged Dictionary*, p. 920
19. Lightman, A. P. *Einsteins Dreams* (New York: Pantheon Books, 1993), p. 27.
20. Williams, Kit, *Masquerade* (New York: Random House, Inc., 1984).
21. Keeton, W. *Biological Sciences*, 3rd ed. (New York: W.W. Norton & Co., 1980), 1080 pages.
22. Ayensu, E. S. and P. Whitfield, *The Rhythms of Life*, New York: Crown Publishers, 1981.
23. Coleman, R. *Wide Awake at 3:00 A.M. By Choice or by Chance?* (New York: W. H. Freeman and Co., 1986), 195 pages.
24. The word contentment has rhythms of n and t.
25. Bünning, E. *The Physiological Clock*, 3rd ed. (New York: Springer-Verlag New York, Inc., 1973), 167 pages.
26. Pasachoff, Jay M. *Astronomy: From the Earth to the Universe*, 3rd ed. (Saunders College Publishing, Philadelphia, 1986), 584 pages.
27. Daan, S. and J. Aschoff, "Circadian contributions to survival." In *Vertebrate Circadian Systems*, eds. J. Aschoff, S. Daan, and G. Groos, (New York: Springer-Verlag New York, Inc., 1982), pp. 305–21.
28. Bennett, M. *Living Clocks in the Animal World* (Springfield, Illinois: Charles C. Thomas, Pub., 1974), p. 106.
29. I prefer to use the word "regimen" for experimental protocols, and the word "regime" for something more political, but not all investigators use my distinction.
30. Wever, R. *The Circadian System of Man, Results of Experiments Under Temporal Isolation* (Springer-Verlag, New York, 1979).
31. Johnson, M. S. "Effect of continuous light on periodic spontaneous activity of white-footed mice (*Peromyscus*)." *J. Exp. Zool.*, 82, 315–28. Johnson concluded the mice possessed a durable self-winding and self-regulating clock.
32. Nelson, R. J. *An Introduction to Behavioral Endocrinology*, (Sunderland, MA: Sinauer Associates, Inc., Pubs., 1995), p. 394.
33. Enright, J. "The search for rhythmicity in biological time series," *J. Theoretical Biology*, 8 (1957), 4226–468.
34. Blackman, R. and J. Tukey, *The Measurement of Power Spectra* (New York: Dover Publications, Inc., 1958), 190 pages.
35. Halberg, F., Y. L. Tong, and E. A. Johnson. "Circadian System Phase—An aspect of temporal morphology; Procedures and illustrative examples," in *The Cellular Aspects of Biorhythms, Symposium on Biorhythms* (New York: Springer-Verlag New York, Inc., 1967), pp. 20–48.

36. Pittendrigh, C. "The circadian oscillation in *Drosophila pseudoobscura* pupae: a model for the photoperiodic clock," *Z. Pflanzenphysiol.* 54, (1966), 275–307.

37. Menaker, M. and A. Eskin. "Entrainment of circadian rhythms by sound in Passer domesticus," *Science*, 154 (1966), 1579–81.

38. Binkley, S. and K. Mosher, "Two circadian rhythms in pairs of sparrows," *J. Biological Rhythms*, 3 (1988), 249–54.

39. Davis, F., S. Stice, and M. Menaker, "Activity and reproductive state in the hamster: Independent control by social stimuli and a circadian pacemaker," *Physiol. Behav.*, 40 (1987), 583–90.

40. Aschoff, J., M. Fatranska, and H. Giedke, "Human circadian rhythms in continuous darkness: Entrainment by social cues," *Science*, 171 (1971), 213–15.

41. Winfree, A. T. *The Timing of Biological Clocks* (New York: Scientific American Books, 1987), 200 pages.

42. Miles, L., D. Raynal, and M. Wilson, "Blind man living in normal society has circadian rhythms of 24.9 hours," *Science*, 198 (1977), 421–23. The record is reprinted in Moore-Ede *et al.*, *The Clocks that Time Us* (Harvard University Press, 1982), p. 374.

43. Bünning, E. *The Physiological Clock*, 3rd ed. (New York: Springer-Verlag, Inc., 1973), 167 pages.

44. Aschoff, J. "The phase-angle difference in circadian periodicity," in *Circadian Clocks*, ed. J. Aschoff, (Amsterdam: North-Holland Pub. Co., 1965), pp. 262–76.

45. Dave Samples, personal communication.

46. Kleinhoonte, A. "uber die durch das Licht regulierten autonomen Bewegungen der Canavalia-Blatter," *Archs. neerl. Sci., ser. IIIb*, 5 (1929), 1–110, 181.

47. Binkley, S., S. Klein, and K. Mosher, "Light and dark control circadian phase in sparrows," in *The Pineal Gland, Endocrine Aspects*, eds. G. Brown and S. Wainwright, (New York: Pergamon Press, Inc., 1984), pp. 59–65.

48. Aschoff, J. K., K. Hoffman, H. Pohl, and R. Wever, "Re-entrainment of circadian rhythms after phase-shifts of the Zeitgeber," *Chronobiologia* 2 (1975), 23–78.

49. Pittendrigh, C. "Circadian systems: Entrainment," in *Biological Rhythms*, ed. J. Aschoff, (New York: Plenum Publishing Corp., 1981), pp. 95–124.

50. Pittendrigh, C. and S. Daan, "A functional analysis of circadian pacemakers in nocturnal rodents. IV. Entrainment: Pacemaker as clock," *J. Comp. Physiol.*, 106 (1976), 291–331.

51. Bünning, E. *The Physiological Clock*, 3rd ed. (New York: Springer-Verlag, New York, Inc., 1973), 167 pages.

52. My students measured their average body temperatures as 98.2°F. The human body temperature has a rhythm about one degree lower at night. Swimming in icy water, a human's body temperature is lowered, and therefore affected by environment, as I can attest.

53. Edmonds, S. and N. Adler, "Food and light as entrainers of circadian running activity in the rat," *Physiol. Behav.*, 18 (1977), 915–19; and "The multiplicity of

biological oscillators in the control of circadian running activity in the rat." *Physiol. Behav.*, 18 (1977), 921–30.

54. I once saw a charming misprint that read "free fun."

55. Binkley, S., K. Mosher, and K. Reilly, "Circadian rhythms in house sparrows: Lighting *ad lib*," *Physiol. Behav.*, 31 (1983), 829–83.

56. Eskin, A. "Some properties of the system controlling the circadian activity rhythm of sparrows," in *Biochronometry*, ed. M. Menaker, (Washington, D.C.: National Academy of Sciences, 1971), pp. 55–77.

57. Aschoff, J. "Circadian activity rhythms in chaffinches (*Fringilla coelebs*) under constant conditions," *The Japanese Journal of Physiology*, 16 (1966), 363–70.

58. The modifying effect of light intensity upon period length of circadian rhythms has been called the "circadian rule."

59. Kleitman, N. T. and T. Englemann, "Sleep characteristics of infants," *Journal of Applied Physiology*, 7 (1953), 269–82.

60. The synthesizing activity of an enzyme, such as pineal N-acetyltransferase, can be measured by supplying radioactive substrates (C14-serotonin and acetyl coenzyme A) to puree of pineal glands and isolating the radioactive products (C14-melatonin and C14-N-acetylserotonin) by thin layer chromatography.

61. Reiter, R., B. Richardson, L. Johnson, B. Ferguson, and D. Dinh, "Pineal melatonin rhythm: Reduction in aging Syrian hamsters," *Science*, 210 (1980), 1372–74.

62. Davis, F. "Ontogeny of circadian rhythms," in J. Aschoff, ed., *Biological Rhythms, Handbook of Behavioral Neurobiology*, 4 (1981), 257–74.

63. Pittendrigh, C. and S. Daan, "Circadian oscillations in rodents: A systematic increase in their frequency with age," *Science*, 186 (1974), 548–50.

64. Webb, W. "Twenty-four hour sleep cycling," in *Sleep*, ed. A. Kales (Philadelphia, PA: J. B. Lippincott Co., 1969), pp. 53–65.

65. Enright, J. *The Timing of Sleep and Wakefulness: On the Substructure and Dynamics of the Circadian Pacemakers Underlying the Wake-sleep Cycle* (New York: Springer-Verlag Inc. 1980), 263 pages.

66. DeCoursey, P. "Phase control of activity in a rodent." *Cold Spring Harbor Symposia on Quantitative Biology*, 25 (1960), 49–56.

67. Brown, F., J. W. Hastings, and J. D. Palmer, *The Biological Clock, Two Views* (New York: Academic Press, Inc., 1970), 94 pages.

68. Bünning, E. *The Physiological Clock*, 3rd ed. (New York: Springer-Verlag, Inc., 1973) 167 pages.

69. Davis, F. "Development of the mouse circadian pacemaker: Independence from environmental cycles," *J. Comp. Physiol.* 143 (1981), 527–39.

70. Konopka, R. and S. Benzer, "Clock mutants of *Drosophila melanogaster*," *Proc. Nat. Acad. Sci. USA*, 68 (1971), pp. 173–81.

71. Takahashi, J. S. and M. Hoffman. "Molecular biological clocks," *American Scientist*, 83 (1995), pp. 159–165.

72. Ralph, M. R. and M. Menaker, "A mutation of the circadian system in Golden hamsters," *Science*, 224 (1988), 1225–7.

73. Tee minus tau = phase shift is the rule of entrainment.
74. Binkley, S. and K. Mosher, "Circadian perching in sparrows: Early responses to two light pulses," *Journal of Biological Rhythms*, 2 (1987), 1–11.
75. Bünning, E. *The Physiological Clock*, 3rd ed. (New York: Springer-Verlag, Inc., 1973), 167 pages.
76. Earnest, D. and F. Turek, "Splitting of the circadian rhythm of activity in hamsters: Effects of exposure to constant darkness and subsequent re-exposure to constant light," *J. Comp. Physiol.*, 145 (1982), 405–11.
77. Pittendrigh, C. "Circadian rhythms and the circadian organization of living systems," *Cold Spring Harbor Symposia on Quantitative Biology*, XXV (1960), 159–84.
78. Chandrashekaran, M., A. Johnsson, and W. Engelmann, "Possible dawn and dusk roles of light pulses shifting the phase of a circadian rhythms," *J. Comp. Physiol.*, 82 (1973), 347–56.
79. Aschoff, J. and R. Wever, "The circadian system of man," in *Biological Rhythms*, ed. J. Aschoff, (New York: Plenum Press, 1981), pp. 311–31.
80. C. Eastman has proposed that they can be explained by an alternative phase shift model, in Aschoff *et al.*, *Vertebrate Circadian Systems* (New York: Springer-Verlag, 1982), p. 262.
81. Sweeney, B. M. and J. Hastings, "Effects of temperature upon diurnal rhythms," *Cold Spring Harbor Symposia on Quantitative Biology*, 25 (1960), 87–113.
82. Bünning, E. *The Physiological Clock*, 3rd ed. (New York: Springer-Verlag, Inc., 1973), 167 pages.
83. Sweeney, B. M. and J. Hastings, "Effects of temperature upon diurnal rhythms," *Cold Spring Harbor Symposia on Quantitative Biology*, 25 (1960), 87–113.
84. Bünning, E. *The Physiological Clock*, 3rd ed. (New York: Springer-Verlag, Inc., 1973), 167 pages.
85. Hibernators reduce their body temperature in bouts. Each bout lasts a few days. The bouts recur throughout the winter season, presumably as an energy saving mechanism to permit the hibernators to survive winter weather extremes and food shortages. During the season of hibernation, the animals reduce body temperature so that it is a few degrees above the ambient temperature in their burrows which are called hibernacula. Hibernating bats and ground squirrels awake at a time of day characteristic of their normal activity phase. Their circadian clocks have continued without error during the low body temperature of the hibernation bout.
86. Binkley, S. *The Clockwork Sparrow*, Englewood Cliffs NJ: Prentice-Hall, 1990, p. 90.
87. The solstices are the times when the sun is the furthest from the celestial equator, about June 21 when the sun attains its most northeastern point on the celestial sphere, or near December 22, when it is at its most southern location.
88. Binkley, S. *The Clockwork Sparrow* (Prentice Hall, Englewood Cliffs, NJ, 1990), p. 107.
89. Palmer's viewpoint, in Palmer, J., F. Brown, and L. Edmunds, *An Introduction to Biological Rhythms* (New York: Academic Press, 1976), 375 pages.

90. Emlen, S. "Migration: Orientation and navigation," in Avian Biology, V, eds. D. Farner and J. King, (New York: Academic Press, Inc., 1975), pp. 129–219.

91. Griffin, Donald R. *Bird Migration*. New York: Dover Publications, 1974, 180 pages.

92. Whittick, A. *Symbols Signs and Their Meaning*, London: Leonard Hill Books Limited (Nine Eden Street, N.W.1), 1960, pp. 239–241.

93. Binkley, S. *The Pineal*, Englewood Cliffs NJ: Prentice-Hall, 1988.

94. The pineal gland is also called the *epiphysis cerebri*. In the human being, the pineal gland is in the center of the brain and was therefore supposed to be the location of the soul. In birds, the pineal gland has a long stalk and the body of the gland is located just beneath the skull on the top of the head at the point where the two cerebral hemispheres and the cerebellum meet.

95. Hoffman, R. and R. Reiter, "Pineal gland: Influence on gonads of male hamsters," *Science*, 148 (1965), 643–58.

96. Ebihara, S., T. Marks, D. J. Hudson, and M. Menaker. Genetic control of melatonin synthesis in the pineal gland of the mouse, *Science*, 231 (1986) 491–3.

97. Binkley, Mosher, and Spangler, in Binkley, S. *The Clockwork Sparrow* (Prentice Hall, Englewood Cliffs, NJ, 1990), p. 126.

98. Kleitman, N. and T. Englemann, "Sleep characteristics of infants," *Journal of Applied Physiology*, 7 (1953), 269–82.

99. Jouvet, M. "Neurophysiological and biochemical mechanisms of sleep," in *Sleep*, ed. A. Kales, (Philadelphia, PA: J. B. Lippincott Co., 1969), pp. 89–100.

100. Hadley, M. *Endocrinology* (Englewood Cliffs, NJ: Prentice-Hall, Inc., 1988), p. 440.

101. Genesis 1:16, King James Bible.

102. Palmer, J. "Daily and tidal components in the persistent rhythmic activity of the crab, *Sesarma*," *Nature*, 215 (1967), 64–66.

103. Barnwell, F. "Daily and tidal patterns of activity in individual fiddler crabs (Genus Uca) from the Woods Hole region," *Biol. Bull.* 130 (1966), 1–17.

104. Youthed, G. J. and R. C. Moran, "The lunar day activity rhythm of myrme-leontid larvae." *J. Insect Physiol.*, 15, 1259–71.

105. Reilly, K. and S. Binkley, "The menstrual rhythm," *Psychoneuroendocrinology*, 6 (1981), 181–84.

106. Dewan, E. "On the possibility of a perfect rhythm method of birth control by periodic light stimulation," *American Journal of Obstetrics and Gynecology*, 99 (1967), 1016–19. The study was replicated by M. Lin, "Night light alters menstrual cycles." *Psychiatry Research*, 33 (1990), 135–8.

107. McClintock, M. "Menstrual synchrony and suppression," *Nature*, 229 (1971), 244–45.

108. Roberts, J. S. Seasonal variations in the reflexive release of oxytocin and in the effect of estradiol on the reflex in goats. *Endocrinology*, 89 (1971), 1029.

109. Vivien-Roels, B. and A. Meinel, "Seasonal variation of serotonin in the pineal complex and the lateral eye of *Lampetra planeri* (Cyclostoma, Petromyzon-tidae)," *Gen. Comp. Endocrinol.*, 51 (1983), 313–23.

110. Hibernation is the physiological process by which some mammals adapt to winter cold extremes by lowering their body temperatures. Woodchucks, marmots, hedgehogs, bats, dormice, hamsters, and ground squirrels hibernate. Hibernation can occur in a warm lighted room. They prepare in the fall by consume enormous amounts of food, gonad regression, and reduced activity of most endocrine glands. During hibernation the animals conserve body water. As they enter hibernation, the heart rate decreases, the respiratory rate decreases, and the deep body temperature remains a few degrees above the environmental temperature. They arouse periodically and urinate, defecate, eat, fix the nest, and sleep. Investigators sprinkle sawdust on a hibernator, and if it is gone the next time they look, the animal has aroused.

111. Gwinner, E. *Circannual Rhythms: Endogenous Annual Clocks in the Organization of Seasonal Processes* (New York: Springer-Verlag, New York, Inc., 1986), 154 pages.

112. Rosenthal, N. E. *Seasons of the Mind* (New York: The Guilford Press, 1989), p. 179.

113. Binkley, *Endocrinology* (Harper Collins, New York, 1995).

114. Winfree, A. *The Timing of Biological Clocks* (New York: Scientific American Books, 1987), 200 pages.

115. Sweeney, B. M. "The photosynthetic rhythm in single cells of *Gonyaulax polyedra. Cold Spring Harbor Symp. Quant. Biol.*, 25 (1960), 145–8.

116. When chick pineal fland cells were grown in Petri dishes, the cells dispersed by "crawling" over the dish much like amoebae. In preliminary observations, we thought there was a daily cycle in the crawling activity.

117. Philippens, K., S. Rover, and J. Abicht, "Circadian rhythmic variations of the relative number of binucleated liver cells in rats," *Prog. in Clin. and Biol. Research*, Vol. 59C (1981), 99–108.

118. Thorud, E., O. P. Clausen, and O. D. Laerum, Circadian Rhythms in Cell population kinetics of self-renewing mammalian tissues, in L. Edmunds, ed., *Cell Cycle Clocks*, New York: Marcel Dekker, Inc. 1984, p. 114.

119. Binkley, S. Rhythms in ocular and pineal N-acetyltransferase: A portrait of an enzyme clock, *Comp. Biochem. Phys.*, 75A (1983) 273–6.

120. Crosthwaite, S. K., J. J. Loros, and J. C. Dunlap, Light-induced resetting of a circadian clock is mediated by a rapid increase in frequency transcript, *Cell*, 81 (1995) 1003–12.

121. Loros, J. The molecular basis of the *Neurospora* clock, *Seminars in The Neurosciences*, 7 (1995), pp. 3–13.

122. Barinaga, M. New clock gene cloned, *Science*, 270 (1995) 732–3; M. P. Myers, *et al.* Positional cloning and sequence analysis of the *Drosophila* clock gene, timeless, *Science*, 270 (1995), 805–8; A. Sehgal, Rhythmic expression of *timeless*: A basis for promoting circadian cycles in period gene autoregulation, *Science,* 270 (1995) 808–10; and Gekakis, N. Isolation of timeless by PER protein interaction: Defective interaction between timeless protein and long-period mutant PER, *Science*, 270 (1995), 811–15.

123. Most of the rats were albino rats raised in the laboratory or purchased from suppliers, but the early work done by Richter used many thousands of wild sewer rats captured in all parts of Baltimore. I always thought that the wheel running records obtained by Richter from his wild rats had more impressive rhythms than did those of the domesticated counterparts.

124. Kennaway, D., J. Peek, T. Gilmore, and P. Royles, "Pituitary response to LHRH, LH pulsatility and plasma melatonin and prolactin changes in ewe lambs treated with melatonin implants to delay puberty," *J. Reprod. Fer.*, 78 (1986), 137–48.

125. Poulton, A., J. English, A. Symons, and J. Arendt, "Changes in plasma concentrations of LH, FSH, and prolactin in ewes receiving melatonin and short-photoperiod treatments to induce early onset of breeding activity," *J. Endocr.*, 112 (1987), 103–11.

126. Czeisler, C., G. Richardson, J. Zimmerman, M. Moore-Ede, and E. Weitzman, "Entrainment of human circadian rhythms by light-dark cycles: A reassessment," in *Photocemistry and Photobiology*, 34 (1981), 239–47.

127. Honma, K., S. Honma, and T. Wada, "Entrainment of human circadian rhythms by artificial bright light cycles," *Experientia*, 43 (1987), 572–74.

128. Czeisler, C., J. Allan, S. Strogatz, J. Ronda, R. Sanchez, C. Rios, W. Freitag, G. Richardson, and R. Kronauer, "Bright light resets the human circadian pacemaker independent of the timing of the sleep-wake cycle," *Science*, 2331 (1986), 667–71.

129. Honma, K., S. Honma, and T. Wada, "Phase-dependent shift of free-running human circadian rhythms in response to a single bright light pulse," *Experientia*, 43 (1987), 1205–07.

130. Wever, R., J. Polasek, and C. Wildgruber, "Bright light affects human circadian rhythms," *Pflügers. Arch.*, 396 (1983), 85–87.

131. Aschoff, J., M. Fatranska, and H. Giedke, "Human circadian rhythms in continuous darkness: Entrainment by social cues," *Science*, 171 (1971), 213–15.

132. Meier-Koll, A., U. Hall, U. Hellwig, G. Kott, and V. Meier-Koll, "A biological oscillator system and the development of sleep-waking behavior during early infancy," *Chronobiologia*, 5 (1978), 425–40.

133. Lewy, A. "Effects of light on human melatonin production and the human circadian system," *Prog. Neuro-Psychopharmacol. & Biol. Psychiat.*, 7 (1983), 551–56.

134. Lille, F. and Y. Burnod, Professional activity and physiological rhythms, *Adv. Bio. Psychiat.*, Vol. 11, 1983, pp. 64–71.

135. Aschoff, J. and R. Wever, "The circadian system of man," in *Biological Rhythms*, ed. J. Aschoff, (New York: Plenum Press, 1981), pp. 311–31.

136. Aschoff, J. and R. Wever, "The circadian system of man," in *Biological Rhythms*, ed. J. Aschoff, (New York: Plenum Press, 1981), pp. 311–31.

137. Winfree, A. T. *The Timing of Biological Clocks* (New York: Scientific American Books, 1987), 200 pages.

138. Magnus, G., M. Cavallini, F. Halberg, G. Cornelissen, D. Sutherland, J. Najarian, and W. Hrushesky, "Circadian toxicology of cyclosporin," *Toxicology and Applied Pharmacology*, 77 (1985), 181–85.

139. Lynch, H. J., R. Rivest, and R. Wurtman, "Artificial induction of melatonin rhythms by programmed microinfusion," *Neuroendocrinology*, 31 (1980), 106–11.

140. Cavallini, M., F. Halberg, G. Cornelissen, F. Enrichens, and C. Margarit, "Organ transplantation and broader chronotherapy with implantable pump and computer programs for marker rhythm assessment," *Journal of Controlled Release*, 3 (1986), 3–13.

141. Vernikos-Danellis, J., C. Winget, and J. Belgjan, "The effect of antiemetic medication on human circadian rhythms," in *Chronopharmacology and Chronotherapeutics*, eds. C. Walker, C. Winget, and K. Soliman, (Tallahassee, FL: Florida A & M University, 1981), pp. 401–11.

142. Richter, C. *Biological Clocks in Medicine and Psychiatry* (Springfield, IL: Charles C. Thomas, Publisher, 1965), 109 pages.

143. Wirz-Justice, A. "Antidepressant drugs: Effects on the circadian system, in T. A. Wehr and F. K. Goodwin eds., *Circadian Rhythms in Psychiatry*, Pacific Grove, CA: The Boxwood Press, 1983, pp. 235–259.

144. Sack, D., N. Rosenthal, B. Parry, and T. Wehr, "Biological rhythms in psychiatry," in *Psychopharmacology*, ed. H. Meltzer, (New York: Raven Press, 1987), pp. 669–85. 86% of the patients were female, 69% of the patients had a family history of affective disorder, and 93% had a lifetime diagnosis of bipolar disorder.

145. Wehr, T., F. Jacobsen, D. Sack, J. Arendt, L. Tamarkin, and N. Rosenthal, "Phototherapy of seasonal affective disorder," *Arch. Gen. Psychiatry*, 43 (1986), 870–75.

146. Lewy, A. "Effects of light on human melatonin production and the human circadian system," *Prog. Neuro-Psychopharmacol. & Biol. Psychiat.*, 7 (1983), 551–56.

147. Sherer, M., H. Weingartner, S. James, N. Rosenthal, "Effects of melatonin on performance testing in patients with seasonal affective disorder," *Neuroscience Letters*, 58 (1985), 277–82.

148. Vaughan, G. Melatonin in humans, in R. Reiter, ed., *Pineal Research Reviews*, (New York: Alan R. Liss, Inc., 1984), pp. 141–201.

149. Rosenthal, N., F. Jacobsen, D. Sack, J. Arendt, S. James, B. Parry, and T. Wehr, "Atenolol in seasonal affective disorder: A test of the melatonin hypothesis," *Am. J. Psychiatry*, 145 (1986), 52–56.

150. Ehret, C. F. and L. W. Scanlon, *Overcoming Jet Lag* (New York: Berkley Pub. Corp., 1983), 160 pages.

151. Aschoff, J., K. Hoffman, H. Pohl, and R. Wever, "Re-entrainment of circadian rhythms after phase-shifts of the Zeitgeber," *Chronobiologia*, 2 (1975), 23–78.

152. Binkley, S. and K. Mosher, "Advancing schedules and constant light produce faster resynchronization of circadian rhythms," *Chronobiology International*, 6, (1989), 305–311.

153. This weekend effect might account for the asymmetry effect.
154. Ehret, C. F. and L. W. Scanlon, *Overcoming Jet Lag* (New York: Berkley Pub. Corp., 1983), 160 pages.
155. Oren, Dan A., W. Reich, N. Rosenthal, and T. Wehr, 1993. *How to Beat Jet Lag, A Practical Guide for Air Travelers.* Henry Holt and Company, New York, 131 pages.
156. Arendt, J. "Light and melatonin as Zeitgebers in man," *Chronobiology International*, 4 (1987), 2743–82.
157. Turek, F. and S. Losee-Olson, "A benzodiazepine used in the treatment of insomnia phase-shifts the mammalian clock," *Nature*, 321 (1986), 167–68.
158. Richard Bachman, Seabrook Shipyard, Texas.
159. Czeisler, C., M. Moore-Ede, and R. Coleman, "Rotating shift work schedules that disrupt sleep are improved by applying circadian principles," *Science*, 217 (1982), 460–63.
160. We tried to mimic shiftwork in the laboratory using house sparrows and light–dark cycles. Sparrows readily re-entrained to repeated phase shifts of five days of LD 8:165 whether it was advanced or delayed. Re-entrainment was fasted if the sparrows were subjected to advancing rotations with intervening constant light (LL). Binkley, S. and K. Mosher, "Advancing schedules and constant light produce faster resynchronization of circadian rhythms," *Chronobiology International*, 6, (1989), 305–311.
161. Czeisler, C., M. Moore-Ede, and R. Coleman, "Rotating shift work schedules that disrupt sleep are improved by applying circadian principles," *Science*, 217 (1982), 460–63.
162. Wever, R., J. Polasek, and C. Wildgruber, "Bright light affects human circadian rhythms," *Pflügers. Arch.*, 396 (1983), 85–87.
163. Mitler, M. M., M. Carskadon, C. A. Czeisler, W. C. Dement, D. F. Dinges, and R. C. Graeber, Catastrophes, sleep, and public policy: Consensus report, *Sleep* (11) 100–109. This is a report of the findings of a committee formed to review reports on the role of human sleep and time of day in medical and human error catastrophes.
164. Reinberg, A., Y. Motohashi, P. Bourdeleau, P. Andlauer, F. Levi, and A. Bicakova-Rocher, *Eur. J. Appl. Physiol.*, 57 (1988), 15–25.
165. Binkley, S., Tome, Maria Begona, and K. Mosher, "Weekly phase shifts of rhythms self-reported by almost feral human students in the USA and Spain." *Physiol. & Behavior*, 46 (1989), 423–27.
166. Binkley, S., M. B. Tome, D. Crawford, and K. Mosher, "Human daily rhythms measured for one year." *Physiology & Behavior*, 48 (1990) 293–298.
167. Binkley, S. "Wrist activity in a woman: Daily, weekly, menstrual, lunar, annual cycles?" *Physiology & Behavior*, 52 (1992), 411–421.
168. Binkley, S. "Individual, phase, and weekly variations in daily cycles of wrist activity in freeliving humans, *Physiology & Behavior*, 53 (1993), 305–207.
169. Binkley, S. "Wrist Motion rhythm phase shifts in travelers may differ from changes in time zones." *Physiology & Behavior*, 55 (1994), 967–70.

170. Palmer, J., F. Brown, and L. Edmunds, *An Introduction to Biological Rhythms* (New York: Academic Press, 1976), 375 pages.
171. Gundel, A., V. Polyakov, and J. Zully, Circadian rhythms and sleep structure in space (Abstract), *Biological Rhythms, 5th Meeting of the SRBR, Program and Abstracts* (1996), p. 9.
172. Baenninger, R., S. Binkley, and M. Baenninger, "Field observations of yawning and activity in humans." *Physiology & Behavior*, 59 (1996), 421–425.
173. Baenninger, R., S. Binkley, and M. Baenninger, "Field observations of yawning and activity in humans." *Physiology & Behavior*, 59 (1996), 421–425.
174. Wever, R. Circadian rhythmicity of man under the influence of weak electromagnetic fields, in M. Moore-Ede, S. Campbell, and R. Reiter, *Electromagnetic Fields and Circadian Rhythmicity* (Boston: Birkhauser, 1992) p. 127.
175. Sulzaman, F., D. Ellman, C. Fuller, M. Moore-Ede, and G. Wassmer, "*Neurospora* circadian rhythms in space: A reexamination of the endogenous-exogenous question." *Science*, 225 (1984), 232–234.
176. An Owl/Lark Self-Test was developed by James Horne and Olov Ostberg, *International J. Chronobiology* 4 (1976), pp. 97–100. The test is reprinted in L. Lamberg, *Bodyrhythms*, Williams Morrow & Co., Inc. New York, 1994, pp. 50–54. The Owl and Lark Questionnaire is also found in *Wide Awake at 3:00 a.m.: By Choice or By Chance?* by Richard M. Coleman (New York: W. H. Freeman & Co., New York, 1986), pp. 182–4.
177. Not to be confused with Temple Owls, which, much like spies, are everywhere.
178. Webb, W. and M. G. Dube, "Temporal characteristics of sleep," in J. Aschoff, ed., *Biological Rhythms* (Plenum, New York, 1981), p. 502.
179. Dinges, D. F. "Napping patterns and effects in human adults," in *Sleep and Alertness: Chronobiological, Behavioral, and Medical Aspects of Napping*, eds. D. F. Dinges and R. J. Broughton, Raven Press, Ltd., NY, 1989.
180. Lamberg, L. *Bodyrhythms* (New York: William Morrow & Co., Inc. , 1994), pp. 95–110.
181. Abrams, W. B. and R. Berkow, *The Merck Manual of Geriatrics*, (Rahway, NJ: Merck Sharp & Dohme Research Laboratories, 1990), pp. 128–140.
182. This is said to be a misconception.
183. Moore-Ede *et al.*, *The Clocks that Time Us* (Harvard University Press, 1982) p. 364.
184. Other sleep disorders include hypersomnia or too much sleeping, narcolepsy which is a sudden attack of sleep, sleep apnea (pickwickian syndrome in obese persons, or snoring), night terrors, nightmares, and somnambulism. Berkow, R. *Sixteenth Edition. The Merck Manual of Diagnosis and Therapy* (Rahway, NJ: Merck Research Laboratories, 1992), pp. 1444–1451.
185. Lingjaerde, O., O. Bratlid, and T. Hansen, "Insomnia during the dark period in northern Norway," *Acat Psychiat. Scand.*, 71 (1985), 506–12.
186. Mercer, H. *November Night Tales* (New York: Walter Neale, 1928); and *The Well of Monte Corbo* (Doylestown, PA: Bucks County Historic Society, 1930).

187. Arking, Robert. *The Biology of Aging*. (Englewood Cliffs, NJ: Prentice-Hall, Inc. Simon & Schuster, 1991), 420 pages.
188. Stewart, F. M. *The Methuselah Enzyme* (New York: Arbor House, 1970).
189. Pierpaoli, W. and G. J. M. Maestroni. "Melatonin: A principal neuroimmunoregulatory and anti-stress hormone: Its anti-aging effects." *Immunology Letters*, 16 (1987), 355–62.
190. Friend, T. "Go Slow on Melatonin Use, Experts Advise," *USA TODAY*, August 13, 1996.
191. A psychopomp, much like a teacher, conducts souls or spirits from this world to other worlds, such as was the work of Hermes or Charon. Thank you Stephen King, I needed that.
192. While writing the Annotated Bibliography of this book, I noticed that in my copy of *The Psychobiology of Curt Paul Richter* some pages were mysteriously bent. The first fold, made in such a way as I have never done (I use Post-it notes), a vertical fold marked the reprinted paper "On the phenomenon of sudden death in animals and man," about Voodoo death. An long angular fold marked page 323, which showed a graph of swimming time versus water temperature (end point drowning, they lasted 60 hours at 95°F) and none of them were asked to swim in cold water. I shivered, wondering if somehow I'd received the Revenge of Richter's Rats.

INDEX

Index